Equivalency Methods for Environmental Liability

Joshua Lipton · Ece Özdemiroğlu
David Chapman · Jennifer Peers
Editors

Equivalency Methods for Environmental Liability

Assessing Damage and Compensation Under the European Environmental Liability Directive

Springer

Editors
Joshua Lipton
Abt Associates
Boulder, CO
USA

Ece Özdemiroğlu
eftec: Economics for the Environment Consultancy
London
UK

David Chapman
Rocky Mountain Research Station
United States Forest Service
Fort Collins, CO
USA

Jennifer Peers
Abt Associates
Boulder, CO
USA

ISBN 978-90-481-9811-5 ISBN 978-90-481-9812-2 (eBook)
https://doi.org/10.1007/978-90-481-9812-2

Library of Congress Control Number: 2017963011

Printed on acid-free paper

This Springer imprint is published by Springer Nature
The registered company is Springer Science+Business Media B.V.
The registered company address is: Van Godewijckstraat 30, 3311 GX Dordrecht, The Netherlands

Ece Özdemiroğlu: Taylan and Tülin

David Chapman: My wife Hollis

Jennifer Peers: My husband Eric and daughter Allison

Foreword

Directive 2004/35/EC of the European Parliament and of the Council of 21 April 2004 on environmental liability with regard to the prevention and remedying of environmental damage (Environmental Liability Directive; ELD) establishes a framework based on the polluter pays principle to prevent and remedy environmental damage. The polluter pays principle is set out in the Treaty on the Functioning of the European Union (Article 191(2) TFEU). It complements the existing permitting and inspection and civil liability regimes.

The Regulatory Fitness and Performance (REFIT) evaluation in 2016 shows that the ELD is working but that it may be reaching only a small part of its potential in terms of addressing environmental damage. One of the recommendations the evaluation makes is to develop tools and measures for a more even and increasing implementation of the ELD, and an essential part of this is better understanding of equivalency methods.

Realising the importance of these methods, the European Commission supported the research project, Resource Equivalency Methods for Assessing Environmental Damage in the European Union (REMEDE) to produce guidelines and case studies in 2006–2008. The Commission initiated and supported also the development of an ELD training material for competent authorities, business operators, loss adjusters, financial security providers, environmental non-governmental organizations (NGOs) and other stakeholders. That material covered in particular the valuation of environmental damage including equivalency methods. We are delighted that the team has developed this book from that beginning and their learning in implementing and training on the methods since then.

We hope this book will be useful for Competent Authorities, operators, financial security providers, assessment experts and NGOs, and contribute to more widespread implementation of the ELD.

Brussels, Belgium

Hans Lopatta
Policy Officer: Legal Issues
European Commission – Directorate General for Environment
Unit E.4 (Compliance & Better Regulation)

Preface

The history of integrated environmental policy in the European Union has seen a shift from prescriptive to enabling legislation. Early legislation tended to be environmental media or economic sector specific, established technological standards, and included command-and-control instruments that focused on penalising damaging behaviour. Later legislation started to incorporate a more integrated approach by addressing multiple environmental media and economic sectors simultaneously; however, policy instruments remained largely prescriptive. More recent legislation has moved towards a generalised yet integrated regulatory approach that leaves the details of implementation to the Member States. This recent approach is intended to create flexible policies specific to the environmental, economic, and social conditions of each Member State, while maintaining overall coherence at the level of the European Union.

The Environmental Liability Directive (ELD) (2004/35/EC) is an example of this latest type of legislation. Its purpose is to establish a framework of environmental liability based on the 'polluter pays' principle, to prevent and remedy environmental damage. On the one hand, it achieves the objective of flexibility in terms of how it fits within the existing national liability regimes and other legislation—for example, in terms of identifying whether damage is sufficiently significant to trigger ELD and whether financial securities should be required. On the other hand, it prescribes the types of resources and damages that fall within its scope, criteria for measuring environmental damage and selecting the appropriate remediation options, and the approaches for quantifying environmental liability known as equivalency analysis methods. Through references to other legislation such as the Water Framework and Habitats and Wild Birds Directives in its definitions of resources, and through its coverage of all economic sectors (albeit with different levels of liability), the ELD also contributes to improving integration across different environmental and other legislation and policy.

The equivalency analysis methods for damage and remediation assessment, even though employed in environmental analysis for more than two decades in the United States under different statutory regimes, were not applied in the European Union until the ELD. This book is based on our experience with implementing equivalency methods in the United States; the research project Resource Equivalency Methods for Assessing Environmental Damage in the European Union (REMEDE) that focused on the development and application of equivalency methods in the context of the ELD and performed illustrative case studies; and providing training programmes to Member States Competent Authorities, private sector operators and non-governmental organisations. Both REMEDE and training programmes were supported by the European Commission.

Acceptable application of equivalency analyses requires technical knowledge (e.g. natural sciences, economics, and law), data, stakeholder engagement, and sometimes a lengthy and costly negotiation process. Collaboration between all stakeholders can make this process more efficient and result in environmentally beneficial outcomes. It is our hope that increased understanding of equivalency analysis will help foster such collaboration.

Boulder, CO, USA — Joshua Lipton
London, UK — Ece Özdemiroğlu
Fort Collins, CO, USA — David Chapman
Boulder, CO, USA — Jennifer Peers

Acknowledgements

The editors would like to acknowledge the many contributions of the REMEDE consortium and the support of the European Commission. The partners included: eftec (UK) (Coordinator and Economic Lead) and Abt Associates (formerly Stratus Consulting) (USA) (Scientific Lead), as well as: PDRF (the Netherlands) (Legal Lead); Autonomous University of Barcelona (Spain), CREAF (Spain); Ecologic (Germany); Swedish University of Agricultural Sciences (Sweden); Pro-Biodiversity Services (Poland); University of Life Sciences (Norway); Jonathan Cox Associates (UK); Royal Society for Protection of Birds (RSPB); Free University Amsterdam (the Netherlands); University of Southern Bohemia (Czech Republic); and University of Gent (Belgium).

In addition, the editors would like to thank Wendy Clough, Jasper Lipton, and Mary Kay Kozyra, for technical editing; Laura Cross and Maia VanPelt, for innumerable and continuously changing travel arrangements; Christine Teter, for graphics design; Jan-Bart Calewaert, for assistance with the REMEDE toolkit; and all the participants in our in-country trainings for helping us test the approaches and for their hospitality. Those who have made chapter-specific contributions have been acknowledged in the relevant chapters.

Focus of the Book

The focus of this book is on equivalency analysis methods. This book is aimed at all those who need to understand the implementation of the Environmental Liabilities Directive (ELD), and in particular Annex II of the ELD, as well as the concept of compensation under Habitats and Wild Birds and Environmental Impact Assessment Directives. This includes environmental scientists, economists, and lawyers from Member State Competent Authorities, private sector operators, and non-governmental organisations.

The text assists the reader in answering two fundamental questions within the scope of the ELD:

1. How can we assess the damages to natural resources and their associated services?
2. How much of what type of remediation is necessary and sufficient to offset these damages?

The tools presented in this book will also be of use in the context of implementing biodiversity offset and habitat banking systems, which have been tested as policy options by the European Commission and individual Member States.

Book Organisation

Part I provides an overview of the policy and technical background covering the legal context within which equivalency analysis fits underlying concepts of equivalency analysis and terminology.

Part II presents a toolkit that outlines five key steps in an equivalency analysis: initial evaluation (Chap. 3); determining and quantifying damage (Chap. 4); determining and quantifying the benefits from remediation (Chap. 5); scaling the complementary and compensatory remediation actions (Chap. 6); remediation planning, monitoring, and reporting (Chap. 7); and economic valuation (Chap. 8).

Part III presents case study applications drawn from settings in different European Member States to illustrate how equivalency analysis could be implemented in different damage contexts. The case studies cover resource, habitat, and value equivalency approaches and reflect different legal contexts and types of damage (physical, biological, chemical). The case studies are founded on real-world situations and include a gas pipeline construction in Poland; construction of a pan-European road; forest fire in Spain; water abstraction in the United Kingdom; and a hypothetical case illustrating equivalency analysis involving damage to alpine brown trout.

Contents

Part III Case Studies

Contributors

Armonía Borrego Center for Research in Environmental Geography, National Autonomous University of Mexico (CIGA—UNAM), Michoacan, Mexico

Edward H. P. Brans Pels Rijcken & Droogleever Fortuijn, The Hague, The Netherlands

David Chapman Rocky Mountain Research Station, United States Forest Service, Fort Collins, CO, USA

Scott Cole EnviroEconomics Sweden Consultancy, Frösön, Sweden

Jonathan Cox Jonathan Cox Associates, Lymington, Hampshire, UK

Eric English Bear Peak Economics, Boulder, CO, USA

Josep Maria Espelta CREAF, Cerdanyola del Vallès, Spain

Jamie V. Holmes Abt Associates, Boulder, CO, USA

Magdalena Kiejzik-Głowińska Eko-Perspektywa, Gdańsk, Poland

Diana Lane Abt Associates, Boulder, CO, USA

Kate LeJeune Rabouillet, France

Joshua Lipton Abt Associates, Boulder, CO, USA

Roberto Molowny-Horas CREAF, Cerdanyola del Vallès, Spain

Ece Özdemiroğlu eftec, London, UK

Jennifer Peers Abt Associates, Boulder, CO, USA

Pere Riera Institute of Environmental Sciences and Technology, Autonomous University of Barcelona, ICTA, Bellatera, Spain

Zenon Tederko Pro-Biodiversity Business Services Ltd, Warsaw, Poland

Abbreviations and Acronyms

ANOVA	Analysis of Variance Tests
BABE	Bages-Berguedà (forest, Catalonia)
BOD	Biological Oxygen Demand
CERCLA	Comprehensive Environmental Response, Compensation, and Liability Act
CI	Confidence Intervals
CLC	Civil Liability Convention
CREAF	Centre for Ecological Research and Forestry Applications
CSM	Conceptual Site Model
CV	Contingent Valuation
CVar	Compensating Variation
CWA	Clean Water Act
DASSYs	Discounted Atlantic Salmon Service Years
DBH	Diameter at Breast Height
DLV	Discounted Loss Value
DRD	Deliberate Release Directive
DSHaYs	Discounted Service Hectare Years
EA	Environment Agency (England and Wales)
ECJ	European Court of Justice
EIA	Environmental Impact Assessment
EIAD	Environmental Impact Assessment Directive
ELD	Environmental Liability Directive
EUNIS	European Union Nature Information System
EVar	Equivalent Variation
EVRI	Environmental Valuation Reference Inventory
FCS	Favourable Conservation Status
GIS	Geographic Information System
GMOs	Genetically Modified Organisms
HD	Habitats Directive

HEA	Habitat Equivalency Analysis (also known as service-to-service method)
ICC	Institut Cartogràfic de Catalunya
MEA	Millennium Ecosystem Assessment
Ml/d	Megalitres/day
MU	Management Unit
NOAA	National Oceanic and Atmospheric Administration (USA)
NRDAR	Natural Resource Damage Assessment and Restoration
OPA	Oil Pollution Act
PEC	Probable Effect Concentration
PSRPA	Park System Resource Protection Act
REA	Resource Equivalency Analysis (also known as resource-to-resource method)
REMEDE	Resource Equivalency Methods for Assessing Environmental Damage in the European Union
SAC	Special Area of Conservation
SCI	Site of Conservation Importance
SEAD	Strategic Environment Assessment Directive
SPA	Special Protection Area
TEC	Threshold Effect Concentration
TEV	Total Economic Value
VEA	Value Equivalency Analysis (including value-to-value and value-to-cost methods)
WBD	Wild Birds Directive
WFD	Water Framework Directive
WTA	Willingness to Accept Compensation
WTP	Willingness To Pay

Part I
Legal and Technical Overview

Chapter 1
The Environmental Liability Directive: Legal Background and Requirements

Edward H. P. Brans

Abstract The objective of the Environmental Liability Directive (ELD) is to provide a common framework for preventing and remediating certain forms of environmental damage. It complements existing *ex ante* European Union nature conservation regimes such as the Habitats, Wild Birds, and Water Framework Directives and provides guidance on how to assess damage to protected natural habitats and species. This chapter provides a legal analysis of the ELD.

Keywords Legal analysis · Environmental liability directive · Habitats directive Wild birds directive · Water framework directive · Damage assessment Standing · Public natural resources

1.1 Introduction

In April 2004, the European Community legislature adopted Directive 2004/35/CE on environmental liability with regard to the prevention and remediation of environmental damage.[1] The objective of this Directive is to provide a common framework for preventing and remediating certain forms of environmental damage. The Environmental Liability Directive (ELD) complements in this respect existing *ex ante* European Union nature conservation regimes, such as those established by the Habitats Directive (HD) and Wild Birds Directive (WBD).[2] Unlike the ELD, these Directives do not contain provisions which enable Member States in *ex post* situations, to order (certain) persons responsible for causing environmental damage

[1]Directive 2004/35/EC, 21.4.2004, OJ 2004 L 143/56.
[2]Resp. Directive 79/409/EEG, OJ 1979 L 103/1 and Directive 92/43/EEG, OJ 1992 L 206/7.

E. H. P. Brans (✉)
Pels Rijcken & Droogleever Fortuijn, Bezuidenhoutseweg 57,
2594 AC The Hague, The Netherlands
e-mail: ehp.brans@pelsrijcken.nl

J. Lipton et al. (eds.), *Equivalency Methods for Environmental Liability*,
https://doi.org/10.1007/978-90-481-9812-2_1

to remediate such damage or to recover the costs of remediation measures if the Member State took these measures itself.

Member States were required to implement the ELD before the end of April 2007, but not every Member State succeeded in this. As is shown by the European Commission report of 12 October 2010, drafted to comply to Article 14(2) of the ELD (European Commission 2010a), the transposition of the ELD was slow and was only completed by July 2010.[3] This resulted in a number of cases before the European Court of Justice (ECJ).[4]

The 2010 European Commission report as well as other more reports also showed a divergent implementation of the ELD across the European Union.[5] The result thereof is that the ELD is not applied among the Member States in a uniform manner.[6] In addition, the 2010 report revealed that the ELD was seldom applied in the European Union Member States in the period 2007–2010.[7] More recent studies, one of them being a 2016 Commission Report on the application of the ELD (European Commission 2016), show that this has not changed significantly. Between April 2007 and April 2013, 1,245 ELD cases were reported. However, 11 Member States have not reported any ELD cases at all and two Member States account for more than 86% of all reported damage cases (European Commission 2016, p. 3). Thus, the number of annual ELD cases per Member State varies considerably, from 95 to less than 1. Most of these reported cases concern soil pollution (around 50%). Damage to water accounts for 30% and damage to biodiversity for around 20% (European Commission 2016).

To further support the use of the ELD, the European Commission recently developed a work programme aimed at improving the application among Member States of the ELD. The aim of the programme is make the ELD deliver better on its original objectives; i.e. to contribute to a better environment by preserving the natural resources, including biodiversity, in the European Union (European Commission 2017).

So most Member States still have no or only a few ELD cases.[8] There are various explanations for this. One is the misconception that the ELD only applies to

---

[3]See European Commission 2010, p. 3 et seq. See also Stevens and Bolton LLP (2013), Table 1 at p. 32.

[4]See Case C-417/08, Commission v. United Kingdom (2009) ECR 2009 I-00106 (judgement ECJ of 18 June 2009) and Case C-422/08, Commission v. Austria (2009) ECR 2009 I-00107 (judgement ECJ of 18 June 2009).

[5]See e.g. BIO Intelligence Service (2013), p. 21 et seq. and Annex A; Stevens and Bolton LLP (2013), pp. 34–103 and Milieu Ltd. and IUCN (2014), Annex I. See also Goldsmith and Lockhart-Mummery (2013), p. 139 et seq.

[6]Milieu Ltd. and IUCN (2014), p. 78 et seq and BIO Intelligence Service (2014), i.e. Tables 1, 3 and 4.

[7]In 2010, 16 ELD cases were identified and it was estimates that the total number of ELD cases across the EU was in 2010 around 50. See European Commission (2010), pp. 4–5 and 9–10.

[8]See BIO Intelligence Service (2013), p. 96 et seq. See also http://ec.europa.eu/environment/legal/liability/index.htm under the heading 'Member State reports on the experience gained in the application of the Directive'.

the most severe instances of damage to the natural resources covered by the ELD (Milieu Ltd and IUCN 2014, p. 78). In addition, the ELD is considered to be difficult to apply; specifically the part of the ELD that deals with assessing damage and compensation in case protected habitats and species have been impacted by a certain incident.

This chapter serves as an introduction to the ELD. It provides an overview of the scope of the regime and discusses its key elements, such available defences, threshold criteria, standing and remediation options.

1.2 Scope of Application of the Environmental Liability Directive

1.2.1 Strict and Fault-Based Liability

The ELD imposes either a strict or fault-based liability—depending on the type of activity involved—on the operator of an occupational activity for damage to (1) protected species and natural habitats, (2) contamination of land and (3) damage to waters covered by the WFD[9] (provided the damage is above a certain threshold) [see Articles 2(1) and 3].[10]

Operators who undertake an activity that is covered by the European Union legislation listed in Annex III of the ELD, can be held strictly liable for the above three types of harm (for which the overarching term 'environmental damage' is used). The EU legislation listed in Annex III of ELD includes directives concerning Integrated Pollution Prevention and Control, on waste management operations, on the geological storage of carbon dioxide, on the transportation of dangerous substances, and on the direct release of genetically modified organisms into the environment. Most of the activities covered by the listed European Commission legislation can be considered environmentally risky activities.

A fault-based liability is imposed on operators of non-listed occupational activities. These operators can only be held liable for damage to protected species and natural habitats and not for the other types of harm mentioned (provided, naturally, that all requirements listed in the ELD are met).

[9]Directive 2000/60/EC establishing a framework for Community action in the field of water policy, OJ 2000 L327/1.

[10]The European Court of Justice ruled in a decision of 13 July 2017 that air pollution as such does not constitute environmental damage covered by the ELD. However, in case airborne elements cause damage tot water, land and protected species and habitats, such damage could come within the scope of the ELD. See ECJ C-129/6, para 41–46.

1.2.2 Exemptions

There are a number of situations that are exempt from the ELD. For example, environmental damage that arises from an incident that falls within the scope of a number of listed international civil liability conventions is not covered by the ELD, provided the convention is in force in the Member State concerned.[11] An example of such a convention is the International Convention of 27 November 1992 on Civil Liability of Oil Pollution Damage [1992 Civil Liability Convention (CLC)]. This convention, which is in force in most of the Member States, covers environmental damage caused by sea-going vessels constructed or adapted for the carriage of oil in bulk, such as oil tankers. It is in that respect irrelevant that the ELD imposes more comprehensive obligations on polluters with regards to the preventive and remediation measures to be taken in case of environmental damage or the threat thereof than these international conventions (Brans 2001; Oosterveen 2004; Nesterowicz 2007; Carbone et al. 2008).[12] This is even the case if damage has been caused to the nature conservation areas designated under the HD and WBD (Natura 2000 sites).[13] This occurred following the incident with the oil tanker Erika in France in 1999. The tanker spilled over 19,500 tonnes of oil and some 400 km of shoreline were affected, including Natura 2000 sites.

Furthermore, an operator may not be held liable if they prove that the damage was caused by a third party (provided appropriate safety measures were in place), or if he proves that the damage resulted from compliance with an order or instruction from a public authority.[14] The ELD also allows Member States the discretion to exempt an operator from liability where the operator demonstrates that he was not at fault or negligent, and that the environmental damage resulted from an emission or event expressly authorised by the regulatory authority.[15] However, according to a judgement of the ECJ of 1 June 2017, the ELD precludes a provision of national law which excludes, generally and automatically, the application of the ELD due to the mere fact that the environmental damage that was caused, is covered by an authorisation granted under that law.[16]

Apart from this so-called regulatory compliance defence, Member States may also decide to exempt an operator from liability where the operator demonstrates that he was not at fault or negligent, and that the environmental damage caused

---

[11] See Article 4(2–4) and Annex IV and V of the ELD.

[12] A difference between the 1992 CLC and the ELD is that under the 1992 CLC, interim losses are not recoverable. See IOPC Funds (2017), p. 7. Since many international liability conventions use the same damage definition as the 1992 CLC, most likely interim losses are also not recoverable under these regimes.

[13] See further on the difference between these international civil liability conventions and the ELD (Brans 2006), pp. 212–214).

[14] See Article 8(3) of the ELD.

[15] See Article 8(4)(a) of the ELD.

[16] ECJ C-529/15, para. 42.

resulted from an emission or event not considered likely to cause environmental damage according to the state of scientific and technical knowledge at the time the emission was released or the activity took place.[17]

These and other options for exemptions have resulted in an inconsistent transposition of the ELD in the various Member States (Horswill 2009, p. 57; BIO Intelligence Service 2009, pp. 24–36; Weissenbacher 2009, pp. 199–202; Fogleman 2009, pp. 147–176; BIO Intelligence Service 2013, pp. 29–76). Thus, although the ELD undoubtedly is an important step in the harmonisation of environmental liability in the European Union, the ELD has not been implemented in the Member States in a uniform way. It should be noted that this could have an impact on the approach to be taken in case of transboundary incidents and possibly makes it more difficult for the public authorities to successfully deal with such cases.

The ELD also does not apply to environmental damage caused by an emission or incident that took place before 30 April 2007, the date by which the ELD should have been transposed into national law. This is affirmed by the ECJ in its decisions of 9 March 2010 and of 1 June 2017.[18] The court concluded in the first judgement that the ELD does apply to 'damage caused by an emission, event or incident which took place after 30 April 2007 where such damage derives either from activities carried out after that date or activities which were carried out but had not finished before that date'.[19] The court does not provide any further guidance as to how to apply this rule in case of the latter. However, it is likely that if evidence is produced that makes it possible to distinguish between damage or the imminent threat thereof which occurred before and after 30 April 2007, the ELD can be applied to the 'new damage' or the threat thereof.

Finally, the ELD does not apply to environmental damage or the imminent threat if such damage is caused by diffuse pollution [Article 5(5) ELD]. However, if the despite the diffuse nature of the pollution, a causal link can be established between the damage caused or threatened to be caused and one or more operators, the ELD can be applied.[20] Obviously, in such cases this often will be difficult.

[17]See Article 8(4)(b) of the ELD. Taken the wording of Article 8(3) and (4) of the ELD, Member States may decide to apply above exemptions to both occupational activities listed in Annex III and non-listed occupational activities. See further on exceptions and defences, Bergkamp and Bergeijk (2013), pp. 80–94.

[18]ECJ C-378/08 (9 March 2010), para. 38–47. The opinion of Advocate General Koddett in this case includes an interesting expose on the applicable *ratione temporis*. See opinion AG Koddett of 22 October 2009 in ECJ case C-378/08. See also ECJ C-529/25 (1 Juni 2017), para. 21–25.

[19]See ECJ C-378/08, at 41.

[20]Ibidem, at 54.

1.2.3 Unlimited Liability

Liability under the ELD is not limited to a certain ceiling.[21] This does not mean, however, that liability is unlimited. Under the ELD, damages are preferably assessed based on the costs of remediation. However, the monetary value of the natural resources and services impacted is an alternative measure of damages that can be used under the ELD should the 'cost of remediation approach' not be appropriate. The ELD contains a set of guidelines on selecting the most appropriate measures to remedy the environmental damage caused (see Annex II of the ELD). These guidelines have been introduced to, among other things, prevent liable operators from being confronted with disproportionate costly restoration measures or a disproportionate claim. According to these guidelines, only reasonable restoration measures are to be taken to remedy the environmental damage caused, thereby taking into account, among other things, the costs of implementing the various restoration options.[22] The ELD does not define a specific point at which the costs of a certain restoration option become disproportionate.

1.2.4 Natural Resources Covered by the Environmental Liability Directive

As noted earlier, the ELD imposes a strict or fault-based liability, depending on the type of activity involved, for (1) damage to protected species and habitats, (2) contamination of land and (3) damage to waters covered by the WFD. Operators who undertake an activity listed in Annex III of the ELD can be held strictly liable for these three types of harm. Operators of non-listed occupational activities can only be held liable for damage to protected species and natural habitats, and not for damage to the waters covered by the WFD or for the contamination of land.[23]

Neither the ELD nor its preamble explains why an operator is exempt from liability if the damage to waters covered by the WFD or the soil pollution damage is caused by a non-listed activity. This is understandable in cases where damage has been caused to nature conservation areas not covered by the ELD and not brought under the scope of the ELD by the Member States. However, if damage is caused to a nature protection area falling under the scope of the ELD, such as Natura

[21]It is beyond the scope of this chapter, but the fact that the ELD does not limit the financial exposure of an operators to a certain amount, has an effect on the availability insurance products for companies and others. For that and other reasons, the European Commission explored the feasibility of establishing a fund (or sectoral funds) and/or risk-pooling scheme(s). See BIO Intelligence Service et al. (2012).

[22]See para. 1.3.1 of Annex II.

[23]See Article 3(1)(a) and (b).

2000 sites, this choice has striking consequences. One is that if it appears necessary to first clean-up the polluted area before one can start taking remediation measures to bring back the impacted natural resources and services to baseline conditions, Member States have to fall back on their national laws to force the operator concerned to take clean-up measures, or to recover the costs of such measures if the Member State took such measures itself. This might prove to be difficult if the soil protection legislation of the Member State concerned is not adequate. This problem might also arise where an operator of a non-listed activity causes damage to the waters covered by the WFD.

1.2.4.1 Protected Species and Habitats

The scope of the ELD to protected species and habitats is, in principle, limited to the species and natural habitats protected by the HD and WBD. However, Member States have the option to bring additional species and natural habitats under the scope of the ELD. This is only possible if such natural resources are protected by national protection and conservation laws.[24]

What natural resources are protected by the HD and WBD? According to Article 1 of the WBD, the Directive applies to all species of wild birds naturally occurring in the European Union including their eggs, nests, and habitats (Sadeleer De 2005, p. 219). Alternatively, the HD has a different approach, as it provides that measures taken pursuant to the HD 'shall be designed to maintain or restore, at favourable conservation status, natural habitats and species of wild flora and fauna of European Community interest'.[25] Thus not all types of natural habitats and species of wild flora and fauna are covered under the HD, only those of Community interest.

The natural habitat types of Community interest are listed in Annex I of the HD. The list includes about 210 natural habitat types. The habitats concerned are either endangered, have suffered from regression, or constitute outstanding examples of the typical characteristics of one or more of the five following biogeographical regions: Alpine, Atlantic, Continental, Macaronesian, and Mediterranean. Species of Community interest are listed in Annex II and/or Annex IV or V of the HD. Such species are either endangered, vulnerable, rare, or endemic.[26]

In order to fulfil the conservation and biodiversity objectives of both the HD and WBD, Member States are required to designate Special Protection Areas (SPAs)

---

[24] Article 2(3)(c).

[25] See Article 2(2) of the HD.

[26] In accordance with Article 1(d) and (h) of the HD, a distinction is made in Annex I and II between respectively so-called priority natural habitat types and other habitat types of Community interest and priority species and other species of Community interest. With regard to the priority natural habitat types and species 'the Community has particular responsibility in view of the proportion of their natural range which falls within [the territory of the Member States]' (Article 1 of the HD).

and Special Areas of Conservation (SACs).[27] The latter are sites hosting natural habitat types listed in Annex I of the HD and the habitats of the species listed in Annex II of this Directive.[28] The SPAs and SACs together form a European ecological network called Natura 2000. These Natura 2000 sites should enable 'the natural habitat types and the species' concerned to be maintained or, where appropriate, restored to a favourable conservation status'.[29] It is expected that about 10–12% of the territory of the European Union will finally be classified as Natura 2000 sites.

In early drafts of the ELD, the proposed liability regime was to be limited to the natural resources located in Natura 2000 sites. Damage to natural resources located outside the Natura 2000 sites would not have been covered, even if damage was caused to species and/or natural habitats protected under the HD and WBD. The geographical limitation of the draft ELD to Natura 2000 sites was considered by non-governmental organizations and others as a serious restriction to the scope of the regime (Betlem and Brans 2002). In response, the final text of the ELD as adopted was revised to set the liability to cover all natural resources protected by the HD and WBD.

1.2.4.2 Damage to Waters Covered by the Water Framework Directive

The ELD also covers damage to waters, however, only insofar as these waters are covered by the WFD. The WFD establishes a framework for water policy in the European Union based on the principle of integrated river basin management.[30] The environmental objectives of the WFD are defined in Article 4. The main objectives of the WFD are the reduction and prevention of water pollution, the protection of the aquatic environment, and the improvement of aquatic ecosystems. The WFD applies to almost all water resources in the European Union, including inland surface waters, transitional waters, groundwater and marine waters under the jurisdiction of Member States.[31] With regard to marine waters, due to the scope of the WFD, in principle only the waters in the coastal strip of a Member State are covered. This is only different where damage is caused or likely to be caused to European Union protected natural habitats and species (European Commission 2010b). Interestingly, in response to the incident with the Deepwater Horizon platform in the Gulf of Mexico in April 2010, the scope of the ELD has been

---

[27]See further on the classification process and the criteria used to select SPAs and SACs (Sadeleer De 2005, pp. 220–231).

[28]See Article 3(1) of the HD.

[29]Ibid.

[30]Linked to the WFD is a number of so-called 'Daughter Directives', one of one of which is the Groundwater Directive (2006/118/EC).

[31]See Article 1 of the WFD. See further Olazábal (2004, pp. 166–170).

extended and now covers all marine waters under the jurisdiction of Member States, including the Exclusive Economic Zone of Member States.[32]

Member States are currently in the process of implementing the WFD. One of their tasks is to make sure that the WFD-related standards and environmental objectives are met for 'protected areas' (unless otherwise specified in other European Commission Directives) by 2015.[33] These include locations designated for the protection of habitats or species where the maintenance or improvement of the status of water is an important factor in their protection, including relevant Natura 2000 sites designated under the HD and WFD.[34] Other examples of protected areas are the 'bodies of water used for the abstraction of water intended for human consumption' and 'areas designated for the protection of economically significant aquatic species'.[35] Member States may assign protected status to areas that have yet not been designated as such.

It should be noted, that damage to the waters covered will only be recoverable if certain threshold criteria about the significance of damage are met. Furthermore, the ELD excludes operators of non-listed activities from liability for damage to these waters.

1.2.4.3 Soil Pollution

The ELD also covers soil pollution. The specific location and ownership of the contaminated land is not material to the liability regime. However, as noted earlier, this type of damage is only recoverable under the ELD if the land damage is caused by a listed potentially dangerous activity (see Annex III). If this is not the case, the operator will not be liable, at least not under European Union law. In fact, most Member States do have laws for the decontamination of soil pollution (Grimeaud 2001; Seerden and Deketelaere 2000).

As with the other types of damage covered by the ELD, land damage is only recoverable if certain threshold criteria for damage are met. Striking is that despite the focus of the ELD on natural resource, where it concerns soil pollution or land damage these criteria only refer to human health risks and not to ecological risks (see Annex II).

---

[32]See Directive 2013/30/EU of the European Parliament and of the Council of 12 June 2013 on safety of offshore oil and gas operations and amending Directive 2004/35/EC, OJ L 178, 28.6.2013, p. 66–106.

[33]See Article 4(1)(c) of the WFD.

[34]See Article 6 and Annex IV of the WFD. See further Grimeaud (2001, pp. 91–92).

[35]See Article 6 and 7 and Annex IV of the WFD.

1.2.5 A Threshold Approach to Natural Resource Damage

The ELD can only be applied in cases where significant damage is or is threatened to be caused to the natural resources covered. It achieves this by defining a threshold of damage above which the ELD's provisions apply and below which Member States have to fall back on national law. For example, with regard to damage to protected species and natural habitats, the ELD applies only if the damage is of such a nature that it has 'significant adverse effects on reaching or maintaining the favourable conservation status' of the habitats and species concerned (article 2.1(a)). The significance of such effects is to be assessed with reference to the baseline condition, taking account of the criteria set out in Annex I of the ELD.

The term 'conservation status of a natural habitat' is defined in the ELD—the wording is similar to that in the HD—as 'the sum of the influences acting on a natural habitat and its typical species that may affect its long-term natural distribution, structure and functions as well as the long-term survival of its typical species within [...] the European territory of the Member States to which the Treaty applies or the territory of a Member State or the natural range of that habitat'.[36] The conservation status of natural habitats is considered favourable when 'its natural range and areas it covers within that range are stable or increasing, [...] the specific structure and functions which are necessary for its long-term maintenance exist and are likely to continue to exist for the foreseeable future, and the conservation status of its typical species is favourable as defined in [Article 2(4)(b) of the ELD]'.

Conservation status of species means—the wording is, again, similar to that in the HD—'the sum of the influences acting on the species concerned that may affect the long-term distribution and abundance of its populations within [...] the European territory of the Member States to which the Treaty applies or the territory of a Member State or the natural range of that species'.[37] The conservation status of a species is considered favourable when 'population dynamics data on the species concerned indicate that it is maintaining itself on a long-term basis as a viable component of its natural habitats', 'the natural range of the species is neither being reduced nor is likely to be reduced for the foreseeable future', and finally 'there is, and will probably continue to be, a sufficiently large habitat to maintain its populations on a long-term basis'.[38]

A comparable approach is taken with regard to damage to the waters covered by the ELD. Water damage is recoverable if it is of such a nature that it 'adversely affects the ecological, chemical and/or quantitative status and/or ecological potential' of these waters.[39] The WFD provides further guidance on how to interpret this threshold.

[36] See Article 2(4)(a) of the ELD. See also Article 1(e) of the HD.

[37] See Article 2(4)(b) of the ELD. See also Article 1(i) of the HD.

[38] See Article 1(a), (e) and (i) of the HD, See also European Commission (2000, pp. 17–18).

[39] Article 2(b) ELD.

Contaminated land can only be claimed for, as was noted earlier, if the contamination is such that it 'creates a significant risk of human health being adversely affected'.[40] For interpretation of this threshold there currently is no European Union Directive that addresses and provides guidance regarding this type of environmental harm. However, Annex II of the ELD provides some guidance.

1.3 Enforcing the Environmental Liability Directive

The ELD contains provisions—in the event of an imminent threat of, or actual, environmental damage—authorising Competent Authorities to require that preventive and/or remediation measures are taken by the operator. In addition, it imposes a duty on operators who caused the imminent threat of, or actual, environmental damage to not only notify the Competent Authority of the fact that environmental damage has occurred or that an imminent threat of such damage occurring exists, but also to take measures to prevent and/or remediate the environmental damage caused.[41] Articles 5–8 and 11 of the ELD are in that respect the most relevant for the ELD's implementation.

Taken the scope of this book, Article 7(1) of the ELD is particularly relevant. This provision imposes a duty on the relevant operator to identify, in accordance with Annex II of the ELD, the potential remedial measures to make good the environmental damage done and to submit them to the Competent Authority for approval. The Competent Authority then decides which remedial measure is to be implemented.

In addition to requiring the operator to take necessary remedial measures, the Competent Authority may implement the remedial measures itself, as a means of last resort. If the Competent Authority takes the remedial measures, the Competent Authority 'shall recover [...] the costs it has incurred in relation to the [...] remediation actions taken under the Directive'.[42] In addition to the above, the Competent Authority is under a duty to assess the significance of the environmental damage caused by the incident.

1.4 Determination of Remediation Measures

According to Article 7 of the ELD, operators 'shall identify, in accordance with Annex II [of the ELD], the potential remedial measures and submit them to the Competent Authority for its approval'. It is then up to the Competent Authority to

[40]Article 2(1)(c) ELD.

[41]See Article 5–7 and 11 ELD, Some of these provisions are so-called self executing provisions. See for further details Fogleman (2006).

[42]See Article 8(2) ELD.

'decide which remedial measures shall be implemented in accordance with Annex II, and with cooperation of the relevant operator'.

1.4.1 Measure of Damages and the Objective of Remediation Measures

One of the primary objectives of the ELD is to restore damage to the species and natural habitats protected under the HD and WBD and to the waters covered by the WFD.[43] The ELD therefore emphasises restoration and chooses restoration costs as the primary and preferred method to assess damages.[44] However, because it takes time to restore the damaged natural resources to baseline condition—that is the condition the natural resources and services would have been in, had the damage not occurred—the operator will also be held liable for the loss or impairment of natural resources and natural resource services during the restoration period (interim losses).[45] In addition to restoration costs and interim losses, the responsible party can be held liable for the costs of assessing damages as well as the administrative, legal and enforcement costs, the costs of data collection and monitoring and oversight costs.[46]

According to Annex II of the ELD, restoration of damage to waters and protected species and natural habitats is to be achieved by way of so-called primary, complementary, and compensatory remediation measures. It should be noted that the objective of these remediation measures is not only to bring back the damaged natural resources to baseline condition, but also to restore the impaired natural resource services to baseline condition.[47] Natural resource services are defined in the ELD as 'the functions performed by a natural resource for the benefit of another natural resource or the public'.[48] For example, a coastal wetland provides food and nesting habitats for birds and other species, clean water for fish populations, and is important for biodiversity maintenance and for pollution assimilation. Examples of human benefits deriving from coastal wetlands include recreational fishing and boating, beach use, wildlife viewing, hiking, and hunting. This is akin to the ecosystem services approach (e.g. MEA 2005) that has become more commonly known since the drafting of the ELD (see Chap. 2 for more details on ecosystem

[43]Although one of the goals of the ELD is to maintain biodiversity, the ELD focuses primarely on the protection and conservation of the natural resources covered by these nature conservation Directives and not or only indirectly on biodiversity as such. See further Brans and Dongelmans (2014).

[44]See Article 7(1) ELD.

[45]See Article 2(11), (13) and Annex II, para. 1(c) and (d).

[46]See Article 8(2) jo 2(16) of the ELD.

[47]See Article 2(15) and Annex II, para 1(b)–(d) of the ELD.

[48]Article 2(13) ELD. See also paragraph 1(d) of Annex II.

services). It should be noted that the other European Union nature conservation Directives, including the HD, do not refer to the human benefits deriving from the species and natural habitats covered by these Directives.[49] So unlike the ELD, such human benefits do not play a role when considering measures to fulfil the obligations under Article 6(4) of the HD or the obligations under Article 6(2) of the HD.[50] The focus of the HD is entirely on species and natural habitats and not—or only indirectly—on the human services provided by these species and habitats.

1.4.2 Primary, Complementary and Compensatory Remediation Measures

As noted earlier, according to the ELD, damage to waters and protected species and natural habitats is to be restored to baseline condition by way of primary, complementary, and compensatory remediation measures. Primary remediation is defined in Annex II as 'any remedial measure which returns the damaged natural resources and/or impaired services to, or towards, baseline condition'. The focus of these measures is thus on directly restoring the natural resources and services that have been impacted to baseline condition.[51]

Complementary remediation is defined in Annex II as 'any remedial measure taken in relation to natural resources and/or services to compensate for the fact that primary remediation does not result in fully restoring the damaged natural resources and/or services'. The purpose of this type of remediation measures is to provide a similar level of natural resources and/or services at an alternative site, as would have been provided if the damaged site had been returned to its baseline condition.

Because neither of these remediation measures compensate for the loss of ecological and/or human services during the remediation period, compensatory remediation measures also need to be taken to compensate for such interim loss of natural resources and services pending recovery. This compensation often consists of additional improvements to protected natural habitats and species or waters at either the damaged site or at an alternative site.

In order to determine the scale of the complementary and compensatory remediation measures, specific resource equivalency methods are proposed in Annex II,

[49]Such services are also not addressed in the relevant EC guidance documents *Managing Natura 2000 Sites*. The provisions of Article 6 of the HD, Luxembourg 2000 and *Guidance document on Article 6(4) of the 'Habitats Directive' 92/43/EEC*, Brussels 2007.

[50]Article 6(4) HD also uses the term 'compensatory measures'. However, taken the text of this provision, the ECJ's case law and the EC's guidance material that is available on article 6 HD, the term has a different meaning and is not comparable to the one used in the ELD. The measures that need to be taken under this provision of the HD are more likely to be considered primary and complementary remediation measures, to use ELD language.

[51]See in this respect para. 1.1.1 and 1.2.1 of Annex II.

including resource-to-resource, service-to-service, value-to-cost, and value-to-value approaches (see further Chaps. 2–8 of this book).

A few observations about the framework outlined in Annex II are appropriate here. The first concerns the option to take complementary remediation measures at an alternative site. It is noted in Annex II that where possible and appropriate, the alternative site needs to be 'geographically linked to the damaged site', 'taking into account the interests of the affected population'. What is meant by the terms 'geographically linked' and 'affected population' is not specified. It is also not clear whether the latter refers to humans, to non-human species, or both. Both the English and Dutch version of the ELD are unclear on this point. However, since the German version of the ELD uses the term 'betroffenen Bevölkerung', it appears as though the language specifically concerns human interests. Taking this version of the ELD as a starting point, human interests as well as impacted non-human species are to be taken into account when selecting the location of the alternative site. The ELD is also not very clear as to the geographical scope of the term 'affected population' and whether or not the term refers to the local community impacted by the loss or impairment of the natural resources covered by the ELD or, more widely, the community in that region of the given Member State or the nation in general. This is relevant to the application of the ELD as it determines whether or not it is permitted to undertake remediation further away from the place where the damage occurred, not benefiting the impacted local or regional community.

Second, Annex II requires that the alternative site is geographically linked to the damaged site. However, no further guidance is provided regarding the (maximum) allowable distance between the impacted site and the alternative site. According to European Commission (2000), for sites designated under the HD, the site selected for compensation should be located within the biogeographical region concerned (i.e. Alpine, Atlantic, Continental, Macaronesian, or Mediterranean). For sites designated under the WBD, the area selected must be located 'within the same range, migration route or wintering area for bird species […] in the Member States concerned' (European Commission 2007).

Whether or not these guidelines apply to ELD incidents is not entirely clear, but it seems likely. If so, the consequence thereof is that when selecting the location of alternative sites, the focus is not so much on the territory of the Member State where the incident occurred, but on the biogeographical region concerned. It might thus well be that the alternative site that is taken to be the most suitable for remediation is located outside of the territory of the Member State where the damage was caused. To my knowledge, there are no examples whereby remediation measures were taken outside of the Member State where the damage was done.

Finally, the framework of complementary, compensatory, and primary remediation measures does not apply to soil pollution cases. Annex II of the ELD states that the aim of the remediation measures for land damage is to 'ensure, as a minimum, that the relevant contaminants are removed, controlled, contained or diminished so that the contaminated land, taking account of its current use or approved future use at the time of the damage, no longer poses any significant risk of adversely affecting human health'.[52] Interim losses for soil damages are not referred to in the ELD. So, most likely interim losses are not to be considered when selecting the most appropriate measure(s) to remediate land damage.[53] However, to my knowledge that is no case law that affirms this.

1.4.3 Selection of the Most Appropriate Remediation Measures

According to Annex II of the ELD, a reasonable range of remediation options—each consisting of a primary, and if necessary a complementary component, and compensatory component—should be developed.[54] The Competent Authority then evaluates the various options and selects the most appropriate one on the basis of a set of criteria.[55] These criteria include: the costs of implementing the various options; the extent to which each option avoids collateral damage and benefits each damaged natural resource and/or service; the likelihood of success of each option; the length of time it will take under each option to restore the damaged resources and services to baseline condition; the extent to which each option achieves the restoration of the site; and the geographical linkage to the damaged site if measures are taken elsewhere.[56] There is no hierarchy of selection criteria.

The process of identifying, evaluating, and selecting remediation options is also important for determining the extent of the operator's liability. As noted earlier, liability under the ELD is in principle unlimited. Because the ELD does not set a liability limit or a standard or numerical ratio for determining at which point the costs of remediation become unreasonable, the weighing of all of the aforesaid (and other) criteria when selecting the most appropriate remediation options is highly relevant to the potential liable person.

According to Articles 7(2) and 11 of the ELD, the Competent Authority finally decides which of the selected remedial options are to be implemented.[57] It might occur that the remedial measures adopted are less successful than expected. The Competent Authority is in that case entitled to alter the remedial measures previously adopted and/or to decide that additional remedial measures are necessary.[58] However, in such cases, the Competent Authority is required to hear the interested parties before adopting a decision on this issue, especially the operator that is

---

[52]See para. 2 of Annex II of the ELD.

[53]This is affirmed by Article 2(15) of the ELD.

[54]Annex II, para. 1.3.2.

[55]Annex II, para. 1.3.1–1.3.3.

[56]Ibid.

[57]Articles. 7(2) and 11 of the ELD. See also ECJ 9 March 2010, Joined Cases C-379/08 and C-380/08, para. 49–50.

[58]See also ECJ 9 March 2010, Joined Cases C-379/08 and C-380/08, para 51.

required to take such measures and who funds them.[59] Where a Competent Authority considers substantially altering the remedial measures which were chosen, it is required to take into account the criteria of Sect. 1.3.1 of Annex II and to prevent that the operator concerned has to incur 'manifestly disproportionate costs' in comparison to the first remediation option chosen.[60] At the time of writing, there is no case law yet on the latter. It is thus unclear at what point these costs become manifestly disproportionate.

1.5 Outlook

The ELD introduces a complex but interesting and potentially powerful liability regime. Although there have already been a number of incidents in the European Union that are covered by the ELD, there is thus far only limited case law. One reason for this is the slow implementation of the ELD into national law. Another is the degree of awareness by public authorities and others of the ELD and its novel instruments. In addition, many Competent Authorities consider the ELD complex and expensive to apply.[61] The European Commission is aware of this and has taken measures, for instance by making available training materials and is preparing addition measures to support Competent Authorities and others in applying the ELD.[62]

References

Bergkamp and Bergeijk. (2013). Exceptions and defences. In L. Bergkamp & B. J. Goldsmith (Eds.), *The EU environmental liability directive: A commentary* (p. 2013). Oxford: Oxford University Press.

Betlem G. & Brans E. H. P. (2002). *The Future Role of Civil Liability for Environmental Damage in the EU*. Yearbook of European Environmental Law 2002, OUP, Oxford, 183–222.

BIO Intelligence Service. (2009). *Study on the Implementation of the Environmental Liability Directive (ELD) and Related Financial Security Issues*. 1–115.

[59] Ibidem, para. 55. The ECJ considers that such an obligation does not exist in case 'where the urgency of the environmental situation requires immediate action on the part of the environmental authority'. Ibidem, para. 56.

[60] Ibidem, para. 64.

[61] The European Commission funded various studies that investigated the causes of the limited application of the ELD. These studies are accessible via: http://ec.europa.eu/environment/legal/liability/index.htm.

[62] See ELD Training Material, accessible via: http://ec.europa.eu/environment/legal/liability/eld_training.htm, and the ELD Multi-Annual Work Programme 2017–2020, (European Commission 2017).

BIO Intelligence Service et al. (2012). *Study to explore the feasibility of creating a fund to cover environmental liability and losses occurring from industrial accidents*, Final report prepared for European Commission, DG ENV.

BIO Intelligence Service. (2013). *Implementation Challenges and Obstacles of the Environmental Liability Directive*, Final Report prepared for European Commission—DG Environment.

BIO Intelligence Service. (2014). ELD Effectiveness: Scope and Exceptions, Final Report prepared for European Commission—DG Environment.

Brans E. H. P. (2001). *Liability for Damage to Public Natural Resources. Standing, Damage and Damage Assessment*. The Hague/London/New York: Kluwer Law International, Environmental Law and Policy Series.

Brans E. H. P. (2006). *Liability for Damage to Public Natural Resources under the 2004 EC Environmental Liability Directive—Standing, Damage and Damage Assessment,* in Betlem, G., Brans, E. H. P. (Eds.), *Environmental Liability in the EU. The 2004 Directive compared with US and Member State Law*. 2006 London, Cameron May.

Brans E. H. P. & Dongelmans D. H. (2014). *The Supplementary Protocol and the EU Environmental Liability Directive: Similarities and differences.* In Shibata A. (Ed.), International Liability Regime for Biodiversity Damage: The Nagoya-Kuala Lumpur Supplementary Rrotocol, Routledge, 2014, pp. 180–200.

Carbone, S. M., Munari, F., & Di Pepe, L. S. (2008). The Environmental Liability Directive and liability for damage to the marine environment. *Environmental Liability, 2008,* 18–27.

European Commission. (2000). *Managing Natura 2000 Sites*. The provisions of Article 6 of the Habitats Directive, Luxembourg.

European Commission. (2007). *Guidance document on Article 6(4) of the 'Habitats Directive' 92/43/EEC*, Brussels.

European Commission. (2010a). *Report on the Effectiveness of the Environmental Liability Directive*, Brussels 12.10.2010, COM(2010) 581 final.

European Commission. (2010b). Commission Staff Working Document Accompanying the Communication from the Commission on Facing the Challenge of the Safety of Offshore Oil and Gas Activities, Brussels, 12.10.2010, SEC(2010) 1193 final.

European Commission. (2016). *Report from the Commission to the Council and the European Parliament under Article 18(2) of Directive 2004/35/EC on environmental liability with regard to the prevention and remedying of environmental damage*, Brussels 14.4.2016, COM/2016/0204 final.

European Commission. (2017). Multi-Annual ELD Work Programme (MAWP) for the Period 2017–2020. *Making the Environmental Liability Directive more fit for purpose* 2017.

Fogleman, V. (2006). Enforcing the environmental liability directive: Duties powers and self-executing provisions. *Environmental Liability, 14*(4), 127.

Fogleman, V. (2009). The environmental damage regulations; the New Regime. *Environmental Liability, 2009,* 147–176.

Goldsmith, B. J., & Lockhart-Mummery, E. (2013). The ELD's national transposition. In L. Bergkamp & B. J. Goldsmith (Eds.), *The EU Environmental Liabilty Directive: A Commentary*. Oxford: Oxford University Press.

Grimeaud, D. (2001). Reforming EU Water Law: Toward Sustainability, *European Environmental Law Review*, 91–92.

Horswill, P. (2009). The biodiversity provisions of the Environmental Liability Directive – benefits, challenges and implications of inconsistent transposition. *Environmental Liability, 2009,* 57–61.

IOPC Funds. (2017). Guidelines for Presenting Claims for Environmental Damage, IOPC/APR17/4/1, 8 March 2017 (non final).

MEA. (2005). World Resources Institute, *Millennium Ecosystem Assessment, 2005. Ecosystems and Human Well-being: Synthesis*. Washington, DC. 2005.

Milieu Ltd., & IUCN. (2014). Experience gained in the application of ELD biodiversity damage. Final report for the European Commission, DG Environment. Brussels, February 2014.

Nesterowicz, M. A. (2007). *The application of the Environmental Liability Directive to damage caused by pollution from ships*. LMCLQ, 107–118.

Olazábal, C. (2004). *Community Legislation on Water Protection*. In Onida O.: Europe and the Environment, Europa Law Publishing, Groningen.

Oosterveen, W. (2004). *Some Recent Developments Regarding Liability for Damage Resulting from Oil Pollution—From the Perspective of an EU Member State*. In Betlem, G. & Brans, E. H. P. (Eds.), *Environmental Liability in the EU. The 2004 Directive compared with US and Member State Law*. London: Cameron May.

Sadeleer De , N. (2005). *Habitats Conservation in EC Law: From Nature Sanctuaries to Ecological Networks* (pp. 215–252). OUP, Oxford: Yearbook of European Environmental Law.

Seerden, E., & Deketelaere, K. (Eds.). (2000). *Legal Aspects of Soil Pollution and Decontamination in the EU Member States and the United States*. Antwerpen/ Groningen: Intersentia.

Stevens & Bolton LLP (2013). *The Study on Analysis of integrating the ELD into 11 national legal frameworks*, Final Report prepared for the European Commission – DG Environment.

Weissenbacher, M. (2009). Implementation of the environmental liability directive in Austria. *Environmental Liability, 2009,* 199–203.

Chapter 2
Resource Equivalency Methods in the European Union: A 'Toolkit' for Calculating Environmental Liability

Joshua Lipton, Ece Özdemiroğlu, Kate LeJeune and Jennifer Peers

Abstract We developed a methodological toolkit for performing resource equivalency analyses in the European Union, supported by European Commission DG Research and Innovation through the REMEDE project (Resource Equivalency Methods for Assessing Environmental Damage in the European Union). The purpose of the Toolkit is to provide users with an overview of resource equivalency methods in the context of the Environmental Liability Directive, Habitats Directive, Wild Birds Directive, and Environmental Impact Assessment Directive. The Toolkit outlines analytical steps that can be used to assess and remediate different types of environmental damages and incidents covered by these Directives.

Keywords Resource equivalency analysis · Habitat equivalency analysis Value equivalency analysis · Credit · Debit

2.1 Introduction

As part of the REMEDE project (Resource Equivalency Methods for Assessing Environmental Damage in the European Union), supported by the European Commission DG Research and Innovation, we prepared a methodological 'toolkit' for performing resource equivalency analyses. The Toolkit provides users with an

J. Lipton (✉) · J. Peers
Abt Associates, 1881 Ninth St., Suite 201, Boulder, CO 80304, USA
e-mail: Josh_Lipton@abtassoc.com

J. Peers
e-mail: Jennifer_Peers@abtassoc.com

E. Özdemiroğlu
eftec, 4 City Road, London EC1Y 2AA, UK
e-mail: ece@eftec.co.uk

K. LeJeune
Le Moulin du Siffré, 66730 Rabouillet, France
e-mail: klejeune@icloud.com

J. Lipton et al. (eds.), *Equivalency Methods for Environmental Liability*,
https://doi.org/10.1007/978-90-481-9812-2_2

overview of resource equivalency methods in the context of the Environmental Liability Directive (ELD), Habitats Directive (HD), Wild Birds Directive (WBD), and Environmental Impact Assessment Directive (EIAD). The toolkit outlines analytical methods that can be used to assess different types of environmental damages covered by these Directives.

The Toolkit is designed to aid the user in answering two fundamental questions:

1. How are losses of or damages to natural resources and/or services assessed and quantified?
2. How much complementary and compensatory remediation is needed to make the public whole for those losses or damages?

The Toolkit does not, and cannot, contain universally applicable answers to these questions. Neither does it offer best practice guidance that will suit all possible scenarios. The wide range of environmental resources and incidents that can be covered by the relevant Directives, as well as incident-specific complexity, prevents the development of a Toolkit that is universally applicable. The 'right' approach necessarily will depend on context. However, the Toolkit does provide users with a set of approaches that can be applied to a wide array of incidents and settings.

In this Toolkit, we describe and illustrate a number of alternative resource equivalency methods. We note, however, that resource equivalency analysis is only one step in the process of deciding how remediation should most fairly and feasibly proceed. There may be other considerations that Competent Authorities, operators, or other stakeholders may wish to take into account for a given incident. Such site- and incident-specific considerations may also be taken into account in any negotiation toward a final remediation agreement to offset environmental damage.

The Toolkit is intended as a resource, and we provide background information, alternative methodological approaches and suggestions, supporting technical information, and a step-wise process to conducting resource equivalency analyses. The Toolkit should not be used, however, as formal regulatory or legal guidance or interpreted as a set of prescribed methods that must be applied in each case. Rather, the specifics of individual incidents and other case-specific situations should be considered whenever the methods outlined in the Toolkit are applied.

It is emphasised that the Toolkit is not intended to answer the following questions in the context of ELD or the other relevant Directives:

- Is the damage deemed 'significant' and/or have threshold criteria of the ELD been met, as defined in Article 2? Although we address certain scientific and statistic issues related to the concept of 'significance,' determining whether damage to natural resources and/or services is 'significant' in the ELD context ultimately is a legal and policy decision to be addressed by individual Member States in particular cases.
- What primary remediation projects or technologies should be undertaken? The focus of this Toolkit is on methodological approaches to equivalency analysis—a new approach in Europe. By contrast, primary remediation (e.g., contaminated site clean-up) has been regular practice in the European Union for many years.

Quantifying the benefits of primary remediation for purposes of determining residual damage does represent a shift in approach, however, and is an important input to the equivalency framework presented in this Toolkit.
- What is the pre-incident baseline condition? The ELD defines the baseline condition as 'the condition at the time of the damage of the natural resources and services that would have been existed had the environmental damage not occurred, estimated on the basis of the best information available' (Article 2). Determination of baseline conditions is a site- and resource-specific, empirical (and sometimes legal) issue. The Toolkit contains guidance on alternative approaches to quantifying baseline conditions. However, it does not provide information or recommendations regarding specific baseline levels or trends for different natural resources or locations.
- Finally, issues relating to the wider implementation of the ELD (e.g., financial guarantees) and other relevant Directives are not covered in this Toolkit.

The methods described here can be applied in three types of damage scenarios: (1) expected damage, as in the context of the HD, WBD, and EIAD; (2) significant imminent threat of damage in the context of the ELD; or (3) after damage has occurred and has been deemed significant in the context of the ELD.

This means that the Toolkit can be useful in both *ex ante* (cases 1 and 2, above) and *ex post* (case 3) damage situations. *Ex ante* damage refers to damage that is known to occur in the future and where the assessment is undertaken prior to such damage. Such planned activities or projects most likely take place in accordance with the provisions of Article 6 of the HD and the 'imminent threat' referred to in Article 2 of the ELD. *Ex post* damage refers to damages that have already occurred.

It should be noted that the ELD contains the minimum requirements for a liability regime focused on the prevention and remediation of damage to protected natural resources. Member States have the option of maintaining or adopting more stringent legislation and/or specific requirements.

2.2 Overview of Equivalency Analysis

When natural resources are damaged by releases of hazardous chemicals, physical destruction of the environment, or biological agents, actions can be undertaken to remediate the resources and to compensate the public for the loss of those resources and their services during the time that the resources are impaired. Equivalency methods are used to determine the type and amount of complementary and/or compensatory remediation needed to make up for such losses. The conceptual approach of balancing losses with gains is shown in Fig. 2.1.

Losses related to damaged resources can include the quantity of the resource itself (e.g., numbers of organisms, community composition diversity, vegetative cover, water quantity or quality impairments) or a service provided by the resource (e.g., ecological functions or recreational uses). Examples of losses include changes

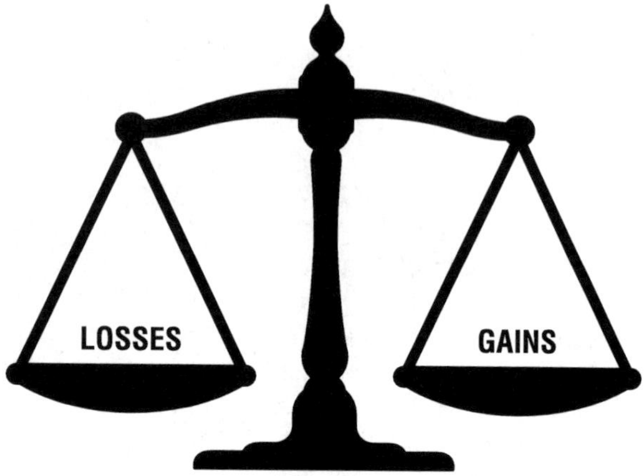

Fig. 2.1 Equivalency analyses—balancing the losses and gains

in the abundance or age-size structure of organisms such as fish, birds, or other wildlife; loss of biodiversity; loss of habitat; reductions in the health, viability, reproductive status, or diversity of organisms; altered community composition or structure; reduced water quantity; impaired water quality; loss of ecosystem functions that contribute to ecological integrity or other services that the public values (such as shoreline protection, flood control, nutrient cycling, water pollutant attenuation, and habitat provision); and loss of use and non-use values placed on the resources, such as fishing, wildlife viewing, or other recreational opportunities, and existence and option values.

Environmental damage can be compensated for through primary, complementary, and compensatory remediation according to the terminology used in the ELD. *Primary remediation*[1] entails actions to reduce or remediate the damage caused by an incident conducted at the site of the incident. Primary remediation generally involves actions such as removal, or clean up, of spilled materials or actions to reduce ongoing discharges of chemicals. Following implementation of primary remediation actions, the damaged natural resources may or may not return to the pre-incident, or baseline, condition (depending on the nature of the incident and the primary remediation actions). The return might be rapid or gradual, depending on the severity of the damage and intensity of the primary remediation actions. In some cases, ecosystem recovery after primary remediation may never reach baseline conditions.

Complementary remediation is needed when recovery after primary remediation will not restore natural resources or services back to baseline conditions. Complementary remediation can be done either at the site of the incident by improving or creating alternative (to the damaged ones) resources or services or at an alternative site by improving natural resources or services of the same or comparable kind. This means that, in some cases, there may be a need to provide

[1]Note that primary remediation is not a component of the Toolkit but is described here for completeness.

enhanced resources or services elsewhere in order to offset a long-term loss expected at the incident site. Complementary remediation can also be appropriate *ex ante* in order to offset resources expected to be lost as a result of planned land use or development, for example, in the context of HD, WBD, and EIAD where such measures are confusingly called 'compensatory remediation' measures.

Compensatory remediation is needed to compensate for losses from the time that damage occurred until recovery to baseline conditions. Such losses are called the 'interim losses'. During the interim loss period, natural resources and the services they provide are diminished or lost. This loss can be offset through remediation of a type and amount of natural resources equivalent to the type and amount lost during the interim period. In such cases, the amount of resources or services lost is calculated in terms of both the quantity of resource loss (e.g., hectares of habitat, fish population reductions) and the duration of the loss.

Equivalency analyses are used to determine the type and amount of resources and services that are lost over time as a result of an environmental damage and the type and amount of actions that are needed to offset the loss. Equivalency analyses take into account the chemical, physical, biological, and, sometimes, social and economic nature of an environmental impact and remediation options.

There are several broad types of equivalency analyses, and depending on the type of analysis, losses and desired remediation benefits can be expressed in different units (or metrics) (see Box 2.1). In this book, we focus on Habitat Equivalency Analysis (HEA; also known as service-to-service), in which losses are expressed in terms of habitat and are offset by remediation of similar habitat, and Resource Equivalency Analysis (REA; also known as resource-to-resource), in which losses are expressed in terms of resource units (such as numbers of fish or birds).

Box 2.1: Hierarchy of Preferred Equivalency Analyses Approaches Identified in Annex II of the Environmental Liability Directive

The ELD provides some level of flexibility for the operators, amongst others, accepting natural recovery as primary remediation as well as intervention, and selecting the type and location of actions. However, it also imposes important constraints related to the hierarchy of the equivalency analyses to be used. The ELD states the following:

> When determining the scale of complementary and compensatory remedial measures, the use of resource-to-resource or service-to-service equivalence approaches shall be considered first. Under these approaches, actions that provide natural resources and/or services of the same type, quality and quantity as those damaged shall be considered first. Where this is not possible, then alternative natural resources and/or services shall be provided. For example, a reduction in quality could be offset by an increase in the quantity of remedial measures (Article 1.2.2, Annex II).
>
> If it is not possible to use the first choice resource-to-resource or service-to-service equivalence approaches, then alternative valuation techniques shall be used. The Competent Authority may prescribe the method, for example monetary valuation, to determine the extent of the necessary complementary and compensatory remedial measures. If valuation of the lost resources and/or services is practicable, but

valuation of the replacement natural resources and/or services cannot be performed within a reasonable time-frame or at a reasonable cost, then the Competent Authority may choose remedial measures whose cost is equivalent to the estimated monetary value of the lost natural resources and/or services (Article 1.2.3, Annex II).

The monetary valuation referred to in Article 1.2.3 implies value-to-value approaches, while 'remedial measures whose cost is equivalent to the estimated monetary value of the lost resources and/or services' refers to value-to-cost approaches.

In summary, the ELD imposes the following hierarchy for resource equivalency approaches to assessing complementary and compensatory remediation:

1. Resource-to-resource
2. Service-to-service methods
3. Value-to-value
4. Value-to-cost

Generally, HEA or REA entails quantifying natural resource losses in terms of a quantity of resource and/or service over time (this loss is often referred to as the 'debit' in an equivalency context), estimation of the quantity of resource or service gain produced by a remediation project (often referred to as the 'credit' in equivalency analysis), and 'scaling' of complementary and compensatory remediation projects (but not primary remediation) to ensure that the total anticipated gain is approximately equal to the calculated loss (i.e. credits $\geq$ debits). The type of environmental damage and opportunities for remediation influence the choice of a specific equivalency approach and the measures of debit and credit.

While (as shown in Box 2.1), the ELD articulates a preference for the use of resource-to-resource or service-to-service scaling approaches such as HEA and REA; approaches using monetary (economic value) units such as value-to-value and value-to-cost might be needed when resource-to-resource or service-to-service approaches are not feasible. Conceptually similar to HEA and REA, the premise of such Value Equivalency Analysis (VEA) is that natural resources provide benefits to the public through the provision of services that can be measured in monetary terms and compensated for following damage in terms of physical resource and service provision.

2.3 Equivalency Analysis: A Brief Historical Overview

Equivalency analysis, including both HEA and REA, was developed in the United States in the early 1990s to quantify environmental liabilities under United States federal laws (Box 2.2).

Box 2.2: Legal Background of Equivalency Analysis in the United States

In the United States, the legal framework for environmental protection includes the common law principles of nuisance, trespass, toxic tort, negligence, public trust, and *parens patriae* (parent of the country), as well as numerous local, state, and federal statutes, regulations, and ordinances. The primary United States federal statutes that address environmental impacts through response and remediation actions are the Comprehensive Environmental Response, Compensation, and Liability Act (CERCLA), the Oil Pollution Act (OPA), the Clean Water Act (CWA), and the Park System Resource Protection Act (PSRPA). These statutes provide for response to oil spills and releases of hazardous substances, clean up of contamination in the environment, physical destruction of resources, and compensation through remediation for public losses caused by releases.

The last of these three provisions, compensation, is the basis for Natural Resource Damage Assessment and Restoration (NRDAR) programs in the United States. These programs ensure that the responsible party pays for the loss of natural resources and the associated remediation costs. When the release (or a response to the threat of a release) of oil or hazardous substances harms public land or other natural resources and when response actions will not fully restore the affected resources, trustees of the public's natural resources can seek compensation from the responsible party to remediate the resources.

In the early 1990s, HEA was applied to sea grass damage claims in the Florida Keys National Marine Sanctuary. Two of these assessments were challenged in United States courts on the applicability of HEA as a reliable method. The conclusions from both court rulings supported the admissibility of HEA as an appropriate method to determine compensatory remediation project scale when the primary category of lost on-site services pertains to the ecological/biological function of an area; when feasible remediation projects are available that provide services of similar type, quality, and comparable value to those that were lost; and when sufficient data to perform the HEA are available or cost effective to collect.

Under the equivalency paradigm, if natural resources or services provided by public resources are lost, the public theoretically can be made whole through replacement of the same or similar natural resources or services. Services provided by natural resources include both human and ecological functions. Examples of services to humans include water and food consumption, fishing, hunting, boating, hiking, bird watching, flood control, shoreline storm protection, and enjoyment of a healthy and functioning natural environment. Services to ecosystems and other ecological resources include basic biological measures (e.g., abundance, diversity, age-size distribution, vegetative structure); habitat for food, shelter, and reproduction; organic carbon and nutrient transfer through the food web; energy transfer through the food web; biodiversity and maintenance of the gene pool; food web and community structure; prevention of the spread of exotic or disruptive species; and natural succession processes. Such functions have been referred to as ecosystem services using the terminology of the Millennium Ecosystem Assessment (MEA 2003, 2005) (see Box 2.3). Under the service-to-service equivalency approach, remediation is scaled so that the service gains provided through remediation equal the service losses caused by the environmental harm.

Box 2.3: What are Ecosystem Services?

The term 'ecosystem services' has come to describe ecosystem structure, function, and human uses. The United Nations MEA, an international work program that is a collaboration of many international organisations (http://www.millenniumassessment.org), formalised the ecosystem services approach by standardising existing concepts and terminology. The MEA defines ecosystem services as the diverse benefits people obtain from ecosystems, which are described as:

- Provisioning services (e.g., food, freshwater, fuel);
- Regulating services (e.g., climate, flood, disease regulation);
- Supporting services (e.g., soil formation, nutrient cycling), and
- Cultural services (e.g., aesthetic, spiritual, recreational).

Ecosystem services in the context of the ELD

Services and natural resources services are the functions performed by a natural resource for the benefit of another natural resource or the public [Article 2(13)]. The ELD thus recognises that natural resources provide services for other elements of nature and for humans alike. Both have to be taken into consideration and accounted for. Resource-to-resource and service-to-service approaches do not express the 'value' of the ecosystem services. They are based on the assumption that the damaged ecosystems *are* sufficiently valuable to justify remediation and deal only with the question of how much remediation is sufficient. Value-to-value and value-to-cost approaches, on the other hand, estimate the *value of the damage* to ecosystem services and compare this to the value and the cost of remediation, respectively.

The first cases in which HEA was used, in the early 1990s, were straightforward incidents involving physical destruction of natural resources, such as where a vessel grounding destroyed sea grass or coral reef habitat. The destruction was clearly delineated, and habitat was essentially eliminated in the footprint of the impact. The date of the destruction was well defined, and estimates of the time required for the habitat to recover naturally and with intervention were estimated based on existing information. The resulting estimate of loss was expressed in lost hectare-years of habitat (i.e. the number of years the area of habitat would be lost). Remediation of an equivalent amount of the same type of habitat was identified as an appropriate compensation for the loss.

These early applications of service-to-service equivalency were innovative developments in two main ways: they extended and formalised the conceptualisation by Freeman (1993) of the environment as an asset that provides a flow of services, and they focused the measure of damages as the scale of remediation projects necessary to compensate for harm over time. One of the key benefits of HEA is that it allows users to bypass the evaluation of monetized economic damages resulting from natural resource damage and to proceed directly to remediation. In addition, HEA explicitly creates a connection between units of services lost because of damage and units of services gained through remediation, when the services provided by proposed remediation actions are of similar type, quality, and value as the services lost.

By the mid-1990s, HEA began to be applied to cases of increasing complexity. In particular, it was applied to cases in which chemical contaminants harmed the environment, but the harm was not so clearly complete as a physical damage that wholly eliminated habitat features. Chemical contaminants can have acute, chronic, or sub-lethal effects on organisms; they can vary over space; and they can persist for long periods of time. These complications required advances in thinking about how to match the scale of the remediation projects to the scale of the damages. In addition, HEA began to be applied in cases where the harm originally began long ago (such as at old mine sites) and where the baseline condition (the condition that would have prevailed had the release or incident in question not taken place) was not directly measured or reported at the time of the release.

As the use of HEA expanded, cases arose where the damage was more appropriately measured in numbers of individuals lost, such as birds or fish, than in habitat units. In such cases, the remediation was scaled to provide equivalent numbers of replacement biological resources, on the theory that the replaced organisms would compensate for the full suite of ecological and human use services lost. This application of resource-to-resource scaling came to be called Resource Equivalency Analysis. The methods of REA are fundamentally the same as for HEA, but the units of quantification differ.

In 1996, the United States National Oceanic and Atmospheric Administration (NOAA) formalised the use of HEA in environmental regulations. Subsequently, resource equivalency methods have been used to scale compensatory remediation at numerous sites in the United States, including in such diverse habitats as Florida coral reefs, salmon habitat in the Northwest, and estuarine wetlands in south Texas.

2.4 Structuring an Equivalency Analysis: Legal Frameworks in the European Union

The equivalency analysis Toolkit presented in this book can be useful when assessing and remediating different types of environmental damage and incidents covered by the ELD and related Directives such as the HD, WBD, Water Framework Directive (WFD), EIAD, and the Strategic Environmental Assessment Directive (SEAD). The appropriate legal framework and Directive should be identified to ensure that the activity that caused the incident is covered and to ensure that the appropriate evaluation criteria are applied when assessing the incident and performing the remediation scaling. In some cases, more than one Directive might apply, depending on the nature of the damage. In such cases, no Directive holds a particular priority *a priori*. Hence, all Directives should be taken into account. Below, we briefly discuss the main aims and coverage of these Directives. Additional details regarding legal aspects of environmental liability regimes are provided in Chap. 1.

2.4.1 The Environmental Liability Directive

The main objective of the ELD is to provide a common framework for remediating (and preventing) environmental damage in the European Union. The ELD imposes both strict- and fault-based liability, depending on the type of activity involved, for damage to the species and habitats covered by the HD and WBD, for contamination of land, and for damage to waters covered by the WFD. As noted in Chap. 1, the ELD was amended in 2013 to specify that damage to water applies to all marine waters of Member States.

Strict liability means that the operator is liable for the damage and loss caused by the operator's acts and omissions regardless of culpability (whether he was at fault or acted negligently is irrelevant). Operators who undertake an activity listed in Annex III of the ELD can be held strictly liable for the three types of harm mentioned above.

Fault-based liability means that an operator can only be held liable for the damage and loss caused by the operator's acts and omissions when at fault or negligent. Operators of non-listed occupational activities can only be held liable for damage to the species and natural habitats covered by the HD and WBD and not for damage to the waters covered by the WFD or for the contamination of land, and only when at fault or negligent. This is illustrated in Fig. 2.2.

The ELD provides several situations that allow Member States to exempt operators from liability. It also should be noted that Member States have certain flexibility with regard to the implementation of the ELD in their national laws and may decide to adopt more stringent rules. Examples include situations in which:

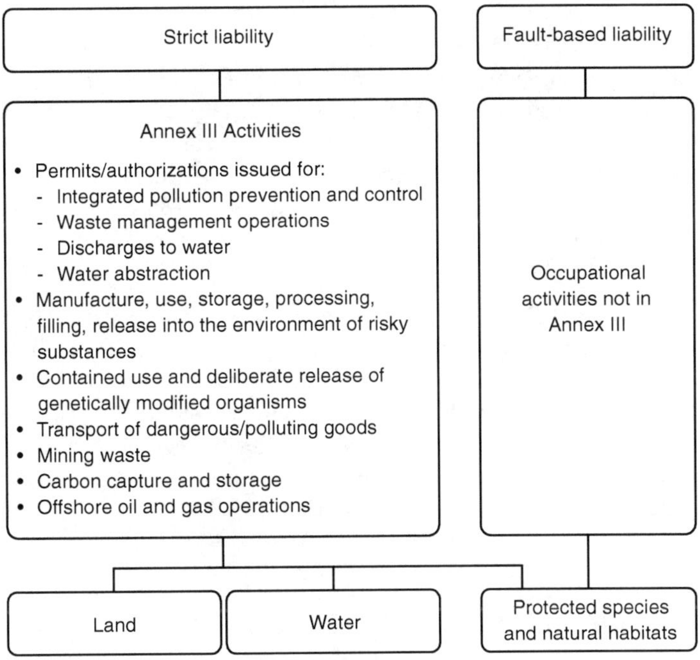

Fig. 2.2 Liability regime and damages covered by the Environmental Liability Directive. *Source* Adapted from Descamps (2005)

- Environmental damage falls within the scope of several listed international civil liability conventions (e.g., the 1992 International Convention on Civil Liability of Oil Pollution Damage), provided the specific convention is in force in the Member State concerned;
- An operator proves that the damage was caused by a third party and occurred despite the fact that appropriate safety measures were in place or if an operator proves that the damage resulted from compliance with a compulsory order or instruction from a public authority, and
- An operator demonstrates that he was not at fault or negligent and that the environmental damage resulted from an emission or event expressly authorised by and fully in accordance with the conditions of an authorisation, or was not considered likely to cause environmental damage according to the state of scientific and technical knowledge at the time the emission was released or the activity took place.

The ELD defines 'damage' as 'a measurable adverse change in a natural resource or measurable impairment of a natural resource service which may occur directly or indirectly' (Article 2). Under the terms of the ELD (Article 2), environmental damage is then defined as:

- Damage to species and natural habitats protected at the Community level by the 1979 WBD or by the 1992 HD where damage means 'significant adverse effects on reaching or maintaining the favourable conservation status';
- Damage to the waters covered by the WFD where damage means significant adverse effects on 'the ecological, chemical or quantitative status or the ecological potential' or 'the environmental statue of … marine waters'[2]; and
- Contamination of the land which creates a significant risk of human health being adversely affected as a result of the direct or indirect introduction in, on or under land of e.g., substances.

Annex II of the ELD provides a 'common framework in order to choose the most appropriate measures to ensure the remedying of environmental damage.' Accordingly, a reasonable range of remediation options, each consisting of a primary and, if necessary, a complementary and compensatory component, should be developed. The Competent Authority then evaluates the various options and selects the most appropriate approach (which may consist of a single option or a combination of actions) on the basis of a set of criteria.

2.4.2 Habitats and Wild Birds Directives

The main objective of both the HD and WBD is the conservation of biodiversity. In that respect, Member States have to take, among other measures, appropriate steps to avoid the deterioration of natural habitats in Natura 2000 sites. They must also avoid the deterioration of the habitats of species and the disturbance of the species for which these sites have been designated.

Both Directives are of relevance for the environmental liability regime that is imposed by the ELD, because they (1) contain important concepts and terminology that the ELD refers to (e.g., the term 'favourable conservation status') and (2) contain information about the biodiversity elements that are covered by the ELD, that is, protected species and habitats.

The HD also has importance in the context of remediation, since Articles 6(3) and (4) of that Directive might result in situations where Member States have to remediate, on an *ex ante* basis, the environmental damage that is caused by the plan or project to be realised and that may have significant effects on Natura 2000 sites. The HD ensures, by means of a preliminary examination—a kind of environmental impact assessment—that a plan or project likely to have a significant effect on a Natura 2000 site is authorised only when it will not adversely affect the integrity of that site. However, adverse effects on the integrity of sites are permitted under certain circumstances and only if Member States take 'all compensatory measures necessary to ensure that the overall coherence of Natura 2000 is protected' [Article 6(4)].

---

[2]As amended by Directive 2013/30/EU of the European Parliament and of the Council of 12 June 2013 on Safety of Offshore Oil and Gas Operations.

It should be emphasised that the compensatory measures of the HD are not the same as the compensatory remediation measures of the ELD. The two terms do not have the same meaning and focus on different types of remediation measures (or at least have different objectives). The purpose of the compensatory measures mentioned in the HD is to provide a similar level of natural resources and services at an alternative site or at a part of the original site not impacted by the project concerned. Therefore, the compensatory measures of the HD are more comparable with the complementary remediation measures of the ELD.

Guidance from the European Commission (2007) specifies that the results of the compensatory measures of the HD need to be operational at the same time as the damage is caused. Hence, no interim losses will be suffered (and have to be compensated for). However, under certain conditions Member States are allowed to take compensatory measures simultaneously to the realisation of the project. In that case, interim losses will be suffered. One way to compensate for such interim losses is to overcompensate for the damage caused (e.g., create a larger habitat than the one that was lost). Given the objectives of the HD and WBD, that is, primarily nature protection and conservation, the extra compensatory measures to be taken, do not aim, at least not directly, at compensating the interim loss of services to humans suffered due to the fact that the compensatory measures were not in place before the Natura 2000 site was damaged. This is an important difference with the ELD, which clearly stipulates that such losses are to be taken into account (see Article 2(13) and paragraph 1(d) of Annex II).

2.4.3 *Water Framework Directive*

The WFD establishes a framework for water policy in the European Union based on the principle of integrated river basin management. As for the HD and WBD, the WFD is important in the context of environmental liability and remediation approaches for three main reasons:

- The scope and extent of coverage provided by the ELD when it comes to water resources is determined by the WFD. In short, the ELD covers damage to waters that are covered by the WFD. The WFD applies to all types of water resources in the European Union, including inland surface waters, transitional waters, coastal waters (up to 1 nautical mile), and groundwater. In 2013, the ELD was amended by Directive 2013/30/EU of the European Parliament and of the Council of 12 June 2013 on Safety of Offshore Oil and Gas Operations to also include any marine waters of Member States.
- The ELD uses concepts from the WFD to describe the damage and set thresholds that trigger the liability regime (e.g., 'ecological, chemical or quantitative status and/or ecological potential'). The ELD refers to the WFD for a full understanding of these terms and concepts (e.g., the definition of ecological status and ecological potential).

- The WFD requires Member States to establish 'programmes for the monitoring of water status in order to establish a coherent and comprehensive overview of water status within each river basin district' [Article 8(1)]. As they are developed, these monitoring programmes will generate data for the determination of baseline conditions for ELD assessments.

2.4.4 Environmental Impact Assessment and Strategic Environmental Assessment Directives

The EIAD requires that Member States ensure that projects likely to have significant effects on the environment because of their nature, size, location, and similar characteristics are subject to development consents and an assessment of their environmental effects. The objective of an Environmental Impact Assessment (EIA) is to identify and describe the environmental impacts of projects and to assess whether prevention or mitigation is appropriate.

As part of the EIA procedure, the project developer is required to submit a description of the measures envisaged to prevent, reduce, and, where possible, offset any significant adverse effects to the environment. Although it is not entirely clear what the objectives of such offsetting are in the context of the EIAD, the Toolkit presented in this book is likely to be of use when determining the nature and extent of the compensatory measures to be taken. In addition, projects subject to an EIA may also require an assessment pursuant to the HD. In these cases, the guidance related to the HD might contain some relevant elements.

Nevertheless, Member States might provide more detailed conditions that need to be fulfilled when identifying the necessary compensatory measures. For example, in the Netherlands, under certain conditions, the anticipated loss of protected natural habitat is to be compensated by the recreation of a new habitat of comparable proportion and with functions comparable to those of the original site. In addition, there is a preference for recreating the new site to be as similar as possible to the impacted site.

Also, the more recently enacted SEAD (Directive 2001/42/EC on the Assessment of the Effects of Certain Plans and Programmes on the Environment) requires that under certain conditions, an environmental assessment is to be undertaken. The Directive stipulates that the assessment report should identify, describe, and evaluate the likely significant effect on the environment of a plan or programme and the reasonable alternatives. Annex I to the Directive specifies that the report should provide information regarding the measures envisaged to reduce and, as fully as possible, offset significant effects on the environment of implementing the plan or program. It does not provide much clarity on the objectives of the measures envisaged to offset the likely significant effects on the environment of a certain plan or programme, nor on how the extent of the measures is to be determined. Also, in this case, there is the possibility that Member States may find it useful to follow some of the approaches described in the Toolkit.

2.5 Equivalency Analysis: A Technical Summary

The technical approach for HEA and REA, as originally presented in a series of published articles (e.g., Peacock 1999; NOAA 2006), is based on the following formulation:

$$\left[\sum_{t=0}^{B} \mathrm{V}^{\mathrm{j}} * \rho_t * \left\{\left(b^j - x_t^j\right)/b^j\right\}\right] * J = \left[\sum_{t=I}^{L} \mathrm{V}^{\mathrm{p}} * \rho_t * \left\{\left(x_t^p - b^p\right)/b^j\right\}\right] * P$$

[____________________________] = [______________________________]

Losses Gains

where t refers to time (in years):

$t = 0$ injury occurs
$t = B$ injured habitat recovers to baseline
$t = C$ time the claim is presented
$t = I$ habitat replacement project begins to provide benefits
$t = L$ habitat replacement project stops providing benefits

and where:

V^j is the annualised per unit value of the benefits provided by the injured habitat (without injury)
V^p is the annualised per unit value of the benefits provided by the replacement habitat
x_t^j is the level of some resource or service metric per hectare provided by the injured habitat at the end of year t
b^j is the baseline (without injury) level of resource or service metric per hectare of the injured habitat
x_t^p is the level of resource or service metric per hectare provided by the replacement habitat at the end of year t
b^p is the initial level of resource or service metric per hectare of the replacement habitat
ρ_t is the discount factor, where $\rho_t = 1/(1+r)^{t-C}$ and r is the discount rate for the time period
J is the number of injured hectares
P is the size in hectares of the replacement project that equates the losses with the gains from remediation

This formulation implies four important points about HEA and REA:

1. Holding all else constant, losses from damages and gains from remediation accrue over different time periods, and ecological benefits gained from remediation conducted in the future are less valuable to the public than ecological benefits available today. To make past, current, and future losses and gains comparable, calculations are made that discount the quantities of natural resources or ecological services from past or future years to present-day terms ('present value'). HEA calculations typically incorporate a discount rate of 3% in the United States, which has the effect of compounding past losses and discounting future losses compared to the present value. In the United Kingdom, the general rule for public sector discounting of a declining rate starting with 3.5% would likely apply (HM Treasury 2003), even though a separate ELD-related discount rate has not been established. The legally required or generally applicable rates should be used in other Member States. Because the remediation typically occurs after the harm, the application of a higher value for a discount rate typically results in more remediation than the use of a lower value. See Chap. 4 (Box 4.4: Discounting) for a more detailed discussion.
2. The quantity of a natural resource or ecological service provided at a damaged area and at a remediation site may be different. In reality, relatively few incidents completely eliminate habitat (or biota), and most remediation actions do not create a completely new and functioning habitat (or biota). In addition, habitat functions are complex, and ecosystem processes are interrelated. To accommodate this complexity, quantification of losses and gains often relies on measurement or estimation of changes based on specified measures, or 'metrics',[3] of a natural resource or ecological service. The metric(s) used must be the same on the loss and gain sides of the equation, and should be useful to discern relative differences in the quality and quantity of natural resources or services.
3. The equivalency approach assumes that the value to society of a given habitat type is essentially constant over time. Alternatively, one might argue that increasing development or climate change may lead to a shortage of some resources or habitat types (e.g., urban wetlands) and thus increase the value of the loss in the future and make its damage more costly today. Resource equivalency does not directly allow for this change in preferences; use of a non-constant discount rate or incorporation of a scarcity index or scalar could indirectly allow for such preferences to be considered.

[3] In some cases, damages may be represented using a single metric for scaling purposes (e.g., hectares of habitat; organism abundance). In other cases, multiple metrics may be employed and the Competent Authority could scale remediation based on the results of a series of equivalency calculations. Finally, compound metrics may also be used in which different measures of resource or service changes are combined into a single metric.

4. All equivalency analyses assume that the public's utility loss can be compensated in the aggregate through remediation or replacement of resources, whatever that may cost (see Flores and Thacher 2002; Zafonte and Hampton 2007).

HEA and REA are appropriate for scaling remediation when (1) a common metric can be defined that reflects the natural resources or services damaged by the impacts and gained through remediation, (2) the landscape contexts of the damaged and remediated habitats are sufficiently similar that the remediation will supply similar natural resources or services (or can reasonably normalized or adjusted using scalars), and (3) sufficient data on HEA and REA input parameters exist, are cost-effective to collect, or can be estimated using professional judgment. As with all models, a lack of input data limits the validity of the outputs. In subsequent chapters, we discuss the issue of estimating natural resource losses and gains, data that may be used to estimate model input parameters, and issues associated with the choice of scaling metrics.

When remediation of the same or similar natural resources or services is not technically feasible (e.g., habitat or organisms of similar type and quality are not available), is undesirable (e.g., if enhancing habitat or number of organisms nearby will increase exposure of wildlife to toxic substances or may cause excessive collateral damage), or is disproportionately or prohibitively expensive, HEA and REA may be less appropriate. In such cases, compensatory actions that provide natural resources or services of different type or quality than those injured may be preferred. In these cases, adjustment factors or VEA may inform the selection and scaling of remediation activities.

2.6 Equivalency Analysis: Debit and Credit

In an equivalency analysis, the *debit* is an expression of the quantity of loss suffered as a result of an environmental damage. The debit is often multidimensional, since an environmental damage can have adverse impacts on many species, habitats, ecosystem functions, and human use and non-use values. In addition, the spatial and temporal extent of the damage and degree of the damage can vary depending on how damage is measured.

Typically in a HEA or REA, one or more measures of loss are defined to serve as indices of keystone resources or services that were damaged. In choosing the measures of debit ('metrics'), an operational assumption often is made that remediation that addresses the chosen metrics may collaterally benefit aspects of the debit that were not specifically treated in the equivalency analysis. For example, if trout populations are selected as the basis for a scaling remediation analysis, it may operationally be assumed that a remediation project that benefits trout would also benefit benthic invertebrates also damaged by the incident. If this is not likely to be the case, the Competent Authority may choose to 'un-couple' the assumed recovery relationship and consider remediation of trout and invertebrates independently. The choice of and use of metrics are discussed further in Chap. 4.

The *credit* in an equivalency analysis is the amount of natural resource or service benefit that will be gained through complementary and compensatory remediation. The number, type, and size of projects are scaled so that the expected amount of benefit generated approximately equals the debit, quantified in terms of the same metric used to quantify the debit.

Ensuring equivalency (scaling) between the debit and credit is conceptually quite simple:

- Quantify the losses (total debits) caused by the damage;
- Determine the amount of benefit (credit) expected per unit of remediation, and
- Divide the total debit by the unit credit to yield the total amount of credits, i.e. remediation needed.

In practice, ecosystem functions are complex, and understanding and quantifying the impacts of a foreseen or unforeseen incident on species, habitats, and/or ecosystem functions can be technically or informationally challenging. In addition, quantifying the benefit that will be provided over time through remediation projects can be informationally limited or technically challenging. Therefore, quantifying the debit and credit typically requires expertise and professional judgment on the part of the equivalency analysis team. Such a team might include biologists, ecologists, toxicologists, chemists, hydrologists, recreation managers, and other environmental specialists, as well as economists and lawyers whose knowledge is relevant to the type of resources and services damaged.

The information and input parameters required to conduct an HEA or REA include the following:

- *Start year*. A start year must be specified for both the debit and the credit side of the model. On the debit side, the start year is the year in which losses began (or are expected to begin) or the year in which the calculation of losses begins. On the credit side, this is the year in which remediation benefits are expected to begin.
- *End year*. An end year can be specified if appropriate. On the debit side, the end year is the year in which losses stop—either through natural recovery or as a result of primary remediation actions. Sometimes there is no expected end year because resources are not expected to recover.[4] On the credit side, this is the last year in which the credit from the remediation project is summed. For some remediation projects, benefit is expected to accrue for the foreseeable future, but in other cases, project life span may be quite limited, particularly if project failure rates (through natural or anthropogenic causes) are considered.

[4]If harm accrues into perpetuity, operators must pay for it. However, because a positive discount rate typically is used, perpetuity often can be approximated by a time frame on the order of 50–100 years (depending on the discount rate used).

- *Base year*. This is the year used for present value calculations. The base year is typically selected as the year in which the analysis is conducted.
- *Spatial extent*. On the debit side of an equivalency analysis, this is the area over which losses have occurred. On the credit side, spatial extent is an expression of the unit area to be remediated. The unit of measure for the credits must be the same as on the debit side to enable equivalency calculations.
- *Degree of damage or service loss*. For HEA, this represents the degree of natural resource or service loss within the spatial extent of damage relative to baseline conditions. The degree of loss can vary over time (as can baseline conditions), and if resource conditions improve over time, natural resources or services may recover to baseline conditions. In a number of cases in the United States, some multi-attribute damage metrics have been expressed in terms of '% service loss,' where the amount of loss can vary from 0 to 100%. It should be emphasised that this concept of a *partial service loss* is not universally accepted. For instance, under certain regulatory regimes, it may be assumed that any damage must be wholly remediated and that, as a practical matter, proportional remediation of a subset of services is neither feasible nor desirable.[5] Similarly, the concept of a single, compound representation of service loss is not universally accepted. Alternative formulations include performing multiple independent, single-metric loss calculations and then scaling remediation based on weighing those different calculations (e.g., selection of the maximum necessary remediation; selecting a mean or some other numerically-weighted combination). For REA, the degree of loss can be expressed in terms of numbers of individuals lost, changes in taxonomic diversity, population reductions, loss of reproductive output or viability (including lost life span or reduced number of young), or other metrics of resource impairment. For VEA, the degree of loss is expressed in monetary terms that reflect the economic value of the loss, that is, individuals' willingness to pay to prevent the loss or willingness to accept compensation to tolerate the loss.
- *Natural resource or service gain*. This is the amount of benefit expected to be derived from implementation of a remediation project. Once a project is implemented, benefits begin to accrue, but full benefits might not be expected until sometime in the future. As with debit calculations, the amount of natural resource or service gain is quantified relative to baseline conditions. Gains (credits) and losses (debits) should be quantified using the same metrics.
- *Baseline conditions*. Baseline conditions are the conditions that would have existed if a damaging incident had not occurred.
- *Metric*. The metric is the unit of measurement of the natural resource or service loss and gain.

---

[5]An analogy that has been used to support this argument is that loss of an arm or a leg in an accident does not represent a 25% loss of 'limb services.' Rather, full compensation would require wholly restoring the injury.

- *Damage or recovery trajectory*. A description of the time course of natural resource or service loss or gain should be preceded and reflect the degradation or recovery rate.
- *Discount (or compound) rate*. To make past, current, and future losses and gains comparable, the quantities of natural resources or services from past or future years are discounted to present-day terms (present value).

Selecting an equivalency method, debit and credit metrics, and appropriate remediation alternatives often can be an iterative process. The analysis team might initially select one equivalency approach and later, when more information is available about the nature of the loss or opportunities for remediation, might decide that another equivalency approach is more likely to allow scaling of an appropriate amount or type of credit.

Other parts of the analysis, namely, selecting a metric(s) to quantify loss and gain, and deciding upon credible and logically consistent approaches (to describe loss and gain trajectories, natural resource or service losses and gains anticipated, and baseline conditions) may also be iterative.

2.7 Typical Steps in an Equivalency Analysis

In general, conducting an equivalency analysis entails five fundamental steps, as described below and in Fig. 2.3. These steps are described in greater detail in Chaps. 3–7.

Step 1: Initial evaluation. This step is performed to determine whether an equivalency analysis should be conducted and, if so, the appropriate scale and content of the analysis.

Step 2: Determining and quantifying damage (the debit). In this step, damaged resources, habitats, and/or services are identified and quantified relative to baseline conditions. The causes of damage are determined. Finally, the benefits of primary remediation are determined and the total debit is quantified.

Step 3: Determining and quantifying the gains from remediation (the credit). Credits are determined by identifying and evaluating potential remediation alternatives and by calculating the benefits that will be gained by implementing complementary or compensatory remediation projects.

Step 4: Scaling complementary and compensatory remediation actions. The final step in the equivalency analysis, per se, is determining the scale or quantity of the remediation project(s) to implement. Scaling is performed so that, over time, the discounted flow of natural resources or services from the remediation projects (credits) is equal to that lost in the impacted area (debits).

Step 5: Monitoring and reporting. After the equivalency analysis is performed and remediation projects are selected and scaled, a remediation plan is prepared that includes project goals, implementation details, engineering plans and designs (if appropriate), and any necessary biological plans and designs. The remediation plan

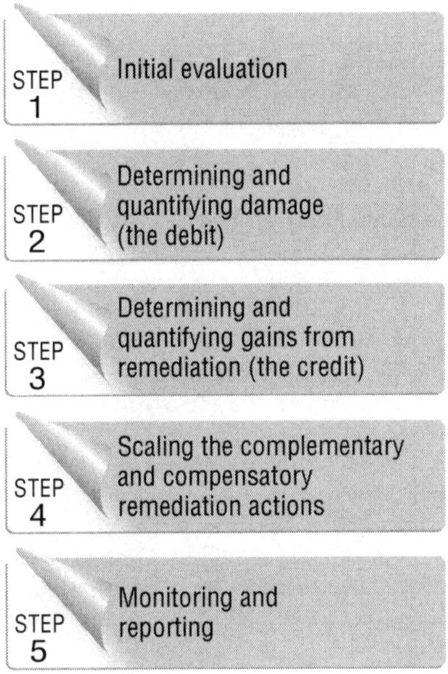

Fig. 2.3 Typical steps in a resource equivalency analysis

also includes procedures and schedules for monitoring the recovery of resources and services following implementation and for evaluating the project's success.

Different levels of assessment may be undertaken in an equivalency analysis. The determination of the appropriate level of detail that should be undertaken typically will be a function of:

- The severity of the incident;
- The degree, extent, and duration of damage;
- The availability of data;
- The ease and cost of additional data collection;
- The degree of precision required for the specific case; and
- Other factors that may be considered by the Competent Authority.

In cases where the spatial and temporal extent and degree of damage are small, and where resources will rapidly return to baseline conditions (with or without primary remediation), equivalency analyses may be undertaken with a limited amount of effort. Such small-scale assessments may rely on readily available data, models, simplifying assumptions and formulas. Where the damage is more complex, likely to cause cascading or persistent adverse effects, cannot be addressed through primary remediation, or simply cannot be addressed quickly, more detailed, comprehensive analyses might be needed. Comprehensive assessments may require data collection and analysis, including design and implementation of field or laboratory studies to understand the extent of the damage or feasibility studies to select appropriate remediation projects or methods.

References

Descamps, H. (2005). Liability for environmental damage in Belgium (Flemish Region) in 2005. *ERA Forum, 6*(3), 401–414.

European Commission. (2007). *Guidance document on Article 6(4) of the 'Habitats Directive' 92/43/EEC*, Brussels.

Flores, N., & Thacher, J. (2002). Money, who needs it? Natural resource damage assessment. *Contemporary Economic Policy, 20*(2), 171–178.

Freeman, M. (1993). *The Measurement of environmental and resource values: Theory and methods*. Washington DC: Resources for the Future.

Millennium Ecosystem Assessment (MEA). (2003). *Ecosystem and human wellbeing: A framework for assessment*. Washington DC: Island Press.

Millennium Ecosystem Assessment (MEA). (2005). *Ecosystems and human wellbeing: Synthesis*. Washington DC: Island Press.

National Oceanic and Atmospheric Administration (NOAA). (2006). *Habitat equivalency analysis: An overview*. Prepared for the National Oceanic and Atmospheric Administration, Damage Assessment and Restoration Program.

Peacock, B. (1999). *Habitat equivalency analysis: Conceptual background and hypothetical example*. Washington DC: National Park Service, Environmental Quality Division.

Treasury, H. M. (2003). *The green book: Appraisal and evaluation in central government*. London: TSO.

Zafonte, M., & Hampton, S. (2007). Exploring welfare implications of resource equivalency analysis in natural resource damage assessments. *Ecological Economics, 61*(1), 134–145.

Part II
Resource Equivalency Analysis: Technical Approaches

Chapter 3
Step 1: Performing an Initial Evaluation

Joshua Lipton and Kate LeJeune

Abstract Before initiating a formal equivalency analysis, the first step in the process typically will entail conducting an initial evaluation of the problem. The objective of the initial evaluation is to determine whether an equivalency analysis should be performed and, if so, the appropriate level of detail for the analysis. Determination of whether an equivalency analysis should be initiated will include a preliminary determination of whether:

- The Environmental Liability Directive (ELD) or other relevant legal framework is applicable to the incident.
- Natural resources are likely to have been damaged as a result of an incident;
- Damages to natural resources are likely to have been significant, as defined by the ELD and relevant national legal frameworks.
- Complementary or compensatory remediation might be needed to offset damages; and
- Equivalency analyses are appropriate to the selection and scaling of remediation.

If primary remediation is undertaken, its timing and anticipated outcomes should be considered during the initial evaluation. Consideration of the influence of primary remediation is integral to developing equivalency analyses. The initial evaluation may also involve identification of the Competent Authorities and their implementation agencies, responsible operators, and other stakeholders that may have a role in the process, either in a public review capacity, or through establishment of cooperative relationships between operators, Competent Authorities, and the affected public. In this chapter, we identify typical elements of an initial evaluation. As each incident is unique, the recommended elements presented here are intended to guide rather than being prescriptive.

J. Lipton (✉)
Abt Associates, 1881 Ninth St., Suite 201, Boulder 80304, CO, USA
e-mail: Josh_Lipton@abtassoc.com

K. LeJeune
Le Moulin du Siffré, 66730 Rabouillet, France
e-mail: klejeune@icloud.com

J. Lipton et al. (eds.), *Equivalency Methods for Environmental Liability*,
https://doi.org/10.1007/978-90-481-9812-2_3

Keywords Initial evaluation · Data availability · Remediation planning
Appropriate scale of assessment

3.1 Introduction

Natural resources can be damaged by various kinds of incidents. An incident could be a chemical release, spill, discharge, or emission; physical impacts or destruction; an introduction or release of a biological agent or entity (e.g., a genetically modified organism or an invasive non-native species); a combination of these; or a by-product, cascading effect, or synergistic effect of a chemical or physical incident.

Once an incident has occurred, the first step in equivalency analysis generally will entail an initial evaluation of the problem. The objective of this initial evaluation is to determine whether an equivalency analysis should be performed and, if so, the appropriate level of detail for the analysis (see Box 3.1, which summarises the key issues that need to be taken into account during this Step of an equivalency analysis).

Box 3.1: Key Issues and Actions in Performing an Initial Evaluation
The fundamental objective of the initial evaluation is to determine whether an equivalency analysis should be performed and, if so, the appropriate level of detail of the analysis. Key questions to answer during the initial evaluation include:

- Are natural resources likely to have been (to be) damaged by an incident covered by the Environmental Liability Directive (ELD) (or other relevant Directives)?
- Are damages likely to be significant (to be determined by the Member States but likely including considerations about extent, severity, and duration of damages)?
- Will primary remediation fully compensate for environmental damages?
- Will complementary or compensatory remediation be needed to offset losses?
- Are services to humans likely to have been or will be affected by the damage?
- What is the appropriate level of detail of the assessment?

The following steps are typically part of the evaluation process (see Fig. 3.1):

1. Identify the relevant Directive(s);
2. Describe the incident;
3. Identify available data;

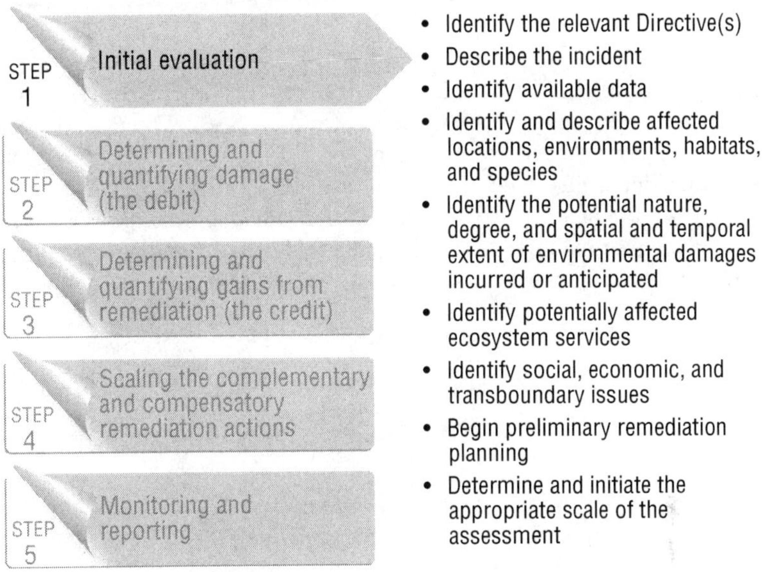

Fig. 3.1 Step 1 of equivalency analysis

4. Identify and describe affected locations, environments, habitats, and species;
5. Identify the potential nature, degree, and spatial and temporal extent of environmental damages incurred or anticipated;
6. Identify potentially affected ecosystem services;
7. Identify social, economic, and transboundary issues;
8. Begin preliminary remediation planning; and
9. Determine and initiate the appropriate scale of the assessment.

To complete these steps, a rapid compilation of available information is typically conducted. These might include conducting a site visit; reviewing literature, databases, reports, and internet sources; and consulting incident responders and experts such as resource managers.

It is important to note that the information collected through the preliminary evaluation may be necessary to determine whether or not a specific incident is covered by the Environmental Liability Directive (ELD) or other relevant legal framework. Therefore, it is not necessary to conclude that an incident is covered by the ELD before conducting a preliminary evaluation.

3.2 Identification of Relevant Directives

The ELD refers to several distinct Directives that describe the resources to which the ELD applies. The appropriate Directive (or enabling national law) should be identified to ensure that the activity that caused the environmental damage is covered. Proper identification also ensures that the appropriate evaluation criteria are applied when the incident is assessed and the remediation scaling is performed. In some cases, more than one Directive can be relevant, depending on the nature of the incident and damage. For details of the Directives relevant in this context, please see Chap. 1.

3.3 Description of the Incident

Details of the incident can inform questions about the potential type and duration of damages, causality and liability, and resource recoverability. The description of the incident should be as detailed as is practical given readily available information. It is important that analysts consider applicable Directives or national legislation when evaluating potential liabilities associated with different types of incidents.

For *ex ante* damages (e.g., in the context of the Environmental Impact Assessment Directive (EIAD), Habitats Directive (HD), and Wild Birds Directive (WBD), plans and designs can provide useful information about the types, timing, severity, and location of anticipated adverse effects. Depending on the situation, it might be prudent to make conservative assumptions about adverse effects. Such assumptions would ensure that an unanticipated outcome does not cause an inordinate difference in the amount of offsetting remediation needed.

For *ex post* damages (e.g., in the context of the ELD), details on the nature, timing, location, and duration of the incident should be compiled. Gathering relevant details could involve preliminary investigatory work. For accidents such as spills and releases, environmental conditions that affect transport and exposure in the environment should be described. In addition, potential adverse effects that might be related to the incident should be identified, and data relevant to determining whether there is a causal link between the incident and potential adverse effects should be identified and considered. During the preliminary evaluation phase, it may be preferable to identify a broad suite of potential adverse effects rather than to risk overly circumscribing potential consequences of an incident.

In identifying the nature of the incident, analysts should attempt to identify those characteristics that may influence the nature and extent of potential adverse effects and will help inform decisions regarding remediation. Types of data and information that might be gathered during the initial evaluation include:

- A detailed description of the release, incident, or project;
- The timing and duration of the event;

- The specific nature of the chemical, physical, or biological stressors associated with the incident;
- Weather conditions;
- Any emergency response actions, primary remediation, or planned mitigation that has already been performed;
- Mapping, tracking, video, and photography/imagery (ground, aerial, or satellite imagery, as appropriate) of the incident, release, or spill;
- Samples of materials that might disperse, dissipate, degrade, denature, or be diluted;
- Supporting environmental data (e.g., temperature, streamflows, pH, dissolved oxygen content, currents, tides, other potential transport vectors);
- Collection of carcasses or data related to transient effects on resources;
- Notes related to scavenging of carcasses;
- Techniques and procedures used to collect ephemeral data;
- Identification of potentially exposed or affected resources and services (including ecological and human services);
- Identification and enumeration of human uses potentially affected by the incident or release;
- Data on the physical, biological, or chemical quality of affected natural resources;
- Data on community ecology relevant to food-chain transfer potential; and
- Baseline information.

Not all of the data or information listed above will be relevant to all incidents. Therefore, it is important that analysts consider the type of preliminary data needed to characterise a specific incident at a given location.

3.4 Preliminary Identification of Available Data

The initial evaluation should include preliminary identification of available data relating to the location of the incident and its effects. This step will assist the Competent Authority in determining the feasibility, ease, and level of detail of a possible assessment. This preliminary identification should include:

- Consideration of the types of available data;
- Quality and quantity of such data;
- Temporal and spatial coverage of data;
- Whether data regarding baseline conditions are available;
- Other information relevant to the identification and description of affected resources and services; and
- Data relevant to determining the degree and extent of the damage, locations, environments, habitats, species, functions, and services.

3.5 Preliminary Identification and Description of Affected Locations, Environments, Habitats, and Species

To aid in the evaluation of potential damages, a preliminary identification of potentially affected locations, environments, habitats, and species should be undertaken. This will facilitate identification of the resources most likely to have been affected or to be at risk from the incident. It might be necessary to evaluate potential resource or habitat scarcity; local or regional importance; and conservation status of potentially affected species, critical habitats, and other local or regional factors that might increase or decrease the likelihood or extent of damage.

Initial evaluation steps might include a site visit; review of incident reports and interviews with any incident responders; review of the literature, databases, and internet sources to identify information relevant to resources (potentially) at risk; and definition of the baseline ecology, biology, and physical attributes of the affected resources. Experts could be contacted for additional information. For example, resource managers often have unpublished data that can be used to characterise baseline conditions and identify potentially affected resources.

3.6 Preliminary Identification of Nature, Degree, and Spatial and Temporal Extent of Environmental Damage

The initial evaluation should include a preliminary identification of the potential nature, degree, and spatial and temporal extent of environmental damage incurred or anticipated. This evaluation could include direct observations (e.g., physical impacts, fish kills, chemical sheens, etc.), descriptions of analogous situations in which damage has been characterised, interviews with incident responders or knowledgeable local observers, accession of satellite or other remote sensing imagery, literature syntheses, comparisons of chemical concentrations with toxicity thresholds, and simple modelling. The preliminary evaluation should attempt to answer the following questions:

- Have resources been exposed to environmental stressors because of the incident?
- What habitats, communities, and species are likely to be at greatest risk of harm?
- Is there direct evidence of damage (e.g., fish kills)?
- What is the nature of potential damage (e.g., mortality, habitat loss, population reductions, contamination that limits productive capacity of habitats)?
- How spatially widespread are potential damages?
- How long might damage persist? Is damage likely to continue into the future?

3.7 Preliminary Identification of Potentially Affected Services

Ecosystem services are the functions performed by a natural resource for the benefit of the public and/or other natural resources. Preliminary identification of damaged services should include evaluation of potentially affected ecosystem services, including ecological services and human use and non-use services. To identify potentially affected ecological functions, the ecology and biology of potentially affected species, communities, habitats, and landscapes should be considered. Examples of ecological services can include:

- Habitat services;
- Maintenance of population dynamics, including consideration of reproductive capacity; maintenance of critical life-stages; age-size distributions; maintenance of necessary reproductive, rearing, foraging, refugia, or other critical habitats;
- Uses of areas as migration corridors;
- Uses of areas as stopover habitats during migration;
- Food chains and nutrient cycling processes that supply energy to sustain populations, habitats, communities, and landscapes;
- Preservation of biodiversity (including at the individual [e.g., genetic], species, population, and habitat levels);
- Alterations in community composition;
- Alterations in landscape dynamics (e.g., edge effects, landscape heterogeneity, thermal properties);
- Loss of the assimilative capacity of wetlands or riparian zones to attenuate contaminants and erosive energy; and
- Reductions in the ability of watersheds to regulate water quality.

In gathering information about potentially affected services, analysts should keep in mind the scarcity or abundance of the services, their regional importance to humans or to the ecosystem, and potential future threats to the area or to resources that provide similar services.

3.8 Preliminary Identification of Social, Economic, and Transboundary Issues

If there are concerns about environmental justice or socially targeted service losses, it might be necessary to obtain existing data that characterises the social and economic landscape of the area under investigation. If the effects of the incident span jurisdictional boundaries, enabling laws, regulations, guidelines, and requirements may vary within the assessment area. Underlying social, economic, and political factors that influence resource use, management, and service flows can also vary. Likewise, if the effects of the incident cross unofficial, but perceived, social or economic geographic boundaries, service losses might vary spatially as a function of these boundaries, in addition to varying with the distribution of contaminants and physical stressors.

3.9 Preliminary Remediation Planning

Remediation planning typically should begin as soon as possible after an incident. Planning includes consideration of whether to conduct primary remediation and, if so, the appropriate nature and extent of primary remediation alternatives. The Toolkit presented in this book does not address primary remediation directly. However, the choice of whether to conduct primary remediation or not and the anticipated benefits of primary remediation are critical when estimating the complementary and compensatory remediation that might be needed. In addition, if the scope of primary remediation can be enhanced to ensure the return of resources and services to baseline conditions (either more completely or more rapidly), then additional remediation might not be needed. Questions to address in the initial evaluation include the following:

(i) *Will primary remediation be conducted?*

Primary remediation is not always feasible or prudent. If conditions at the incident site are endangering human health or safety, certain primary remediation actions may be deemed unacceptable. Similarly, if primary remediation actions are unlikely to substantially benefit the environment or if actions that could be taken are likely to cause substantial collateral damage, primary remediation might not be desirable.

If a decision is made to conduct primary remediation actions, the nature of these actions potentially can be tailored to facilitate the recovery of damaged resources to baseline conditions. If a rapid primary remediation action can remove threats to public health and welfare or the environment and can also return resources rapidly to baseline condition, further complementary or compensatory remediation activities may not be necessary.

By taking the following actions, which can be incorporated into primary remediation goals, the need for complementary and compensatory remediation potentially could be reduced:

- Accelerate recovery to baseline conditions (rather than simply reduce risks to human health or the environment) by reëstablishing, for example:
  - the quantity and quality of surface water flow that occurred before the incident;
  - the quantity (depth) and quality (nutrient cycling ability, nutrient availability; water storage capacity) of soil that was present before the incident;
  - baseline vegetation community composition and structure;

  - components of the food web that support fish and wildlife, such as invertebrate communities essential to insectivorous fish and wildlife, and small mammal communities essential for raptors and carnivorous mammals; and
  - habitats that might have been degraded as a result of damages caused by the incident.

- Reestablish access to the recreational services formerly provided by the area.
- Accelerate the recovery of resource uses, such as fishing or beach-going.
- Reestablish access to commercial services provided by the resources.

Additional primary remediation actions that might be taken to address ecosystem services might include the following:

- Remove contaminated soils to restore baseline soil chemistry (rather than remove contaminants to a clean-up level that differs from baseline) to ensure that soil biota, plants, and soil processes are not adversely affected by residual contamination above baseline concentrations;
- Regrade, recontour, and revegetate with native species to accelerate natural recovery after disturbance related to primary remediation actions; and
- Enhance aquatic habitat through riparian vegetation planting or in-stream work in order to restore a system to its baseline physical condition, or better, after primary remediation actions.

(ii) *Will primary remediation restore baseline conditions quickly?*

If primary remediation actions are expected to restore resources to baseline conditions quickly and completely, there may be no need to consider complementary or compensatory remediation actions. If it is clear that the cost (monetary or otherwise) of estimating the marginal benefit of additional remediation (through equivalency analysis) will exceed the benefit to be gained from the additional remediation, a decision should be made early on whether further analysis of damage and remediation is necessary.

(iii) *Are complementary or compensatory remediation actions appropriate and feasible?*

When primary remediation is not expected to restore baseline conditions quickly and fully, additional off-site remediation (either complementary or compensatory or both) may be needed to return resources and services to baseline condition and/or compensate for interim damages. This situation may arise when:

- Primary remediation, even if designed to restore baseline conditions, will not be conducted until some time in the future;
- Primary remediation, even if designed to restore baseline conditions, will involve actions that take a long time to complete;
- Primary remediation, even if designed to restore baseline conditions, will entail a lengthy recovery period; or

- Primary remediation will not result in a complete recovery of resources/services to baseline conditions.

In the above cases, compensatory remediation may be needed to offset the interim losses that would occur between the time of the incident and the time that baseline conditions are restored. In addition, in the latter situation, complementary remediation is needed to offset the difference between conditions at the site following completion of the primary remediation and baseline conditions. As a practical matter, remediation projects that are implemented to complement and compensate often can be one and the same, particularly if the services to be replaced are similar.

(iv) *Should economic valuation be considered?*

If the lost resources and services are not amenable to any type of remediation, either on- or off-site, an economic valuation might be appropriate. This could occur in cases where the lost resources and services are particularly unique and irreplaceable or where restoring them is technologically infeasible or cost-prohibitive. These economic approaches might be useful when equating public values of loss with gain of different types of resources and/or services ('out-of-kind') or when determining compensatory payments.

For example, monetary compensation for damages might be used to benefit the environment in a way that is not directly related to the losses but is valued equally by the affected public nonetheless. Note that this is not simply money changing hands, it is compensation payment used for environmental improvements that are not necessary directly comparable to the damaged resources or services. Surveys and assessments of individuals' values might be more informative and the resulting out-of-kind improvements more desirable than excessively costly or risky in-kind resource or service remediation projects.

Economic methods might also be preferable if a unique environment that has no equivalent in the area is damaged or an area of such an extent or location that equivalent remediation may be disproportionately costly or even impossible (e.g., sufficiently large areas of the relevant habitat may not exist within a given area) is damaged. Similarly, complementary and compensatory restoration of abundant resources or services might be undesirable. However, funding for a smaller out-of kind project to compensate for a scaled loss might be highly valued by the public. In cases where the replacement project is out-of-kind or ecologically-based adjustment factors are not readily available, economic valuation methods might be one way to equate public values of the loss and gain.

3.10 Initiating and Determining the Appropriate Scale of the Assessment

Based on the preliminary evaluation described above, additional work to scope complementary and compensatory remediation using an equivalency approach may be needed and warranted if the following conditions apply:

- An incident covered by the ELD or related Directives and/or Member State-specific frameworks/regulations has occurred or may occur (including the 'imminent threat' conditions mentioned in the ELD);
- The quantity and concentration of contaminants released or the degree of physical damage are sufficient to potentially cause harm to natural resources;
- Natural resources, or services provided by the natural resources, are potentially damaged;
- Primary remediation actions will not adequately remedy the harm resulting from the incident;
- Opportunities potentially exist off-site to conduct complementary and compensatory remediation projects; and
- Data necessary to quantify damages and conduct remediation planning and scaling are available, can be collected at reasonable cost, can be modeled, or can be reasonably estimated.

Next, a decision should be made about the level of effort to expend in performing the assessment. Depending on the degree, severity, duration, and extent of impacts; the sensitivity or scarcity of affected resources and services; and other transboundary, cultural, and political issues, the assessment could be expedited by using only existing data or models or extended by conducting additional data collection and analysis. Also, uncertainty in the different steps of the equivalency analysis will need to be considered as early as possible.

Chapter 4
Step 2: Determining and Quantifying Environmental Damage

Joshua Lipton and Kate LeJeune

Abstract The purpose of damage determination and quantification is to quantify the lost resources and/or resource services that should be offset by remediation projects. Damage determination and quantification might involve studies to determine the causes, degree, spatial and temporal extent, and nature of damages. In other cases, existing data and/or models may suffice. Damage studies should be designed to produce scientifically rigorous, high quality data and to answer questions relevant to the equivalency analysis. Studies should not be designed primarily to answer questions that are of purely scientific interest. However, analysts should not hesitate to conduct investigations of an exacting scientific nature, because without correct quantitative information about damages, equivalency analysis is unlikely to provide for the 'right' amount of remediation.

Keywords Environmental damage · Debit · Interim loss · Primary remediation Metrics

4.1 Introduction

The second step in performing an equivalency analysis involves the determination and quantification of environmental damage caused by the incident in question. Key elements of this portion of the analysis are to (Box 4.1 and Fig. 4.1):

- Identify damaged resources, habitats, and services;
- Determine the causes of damage;

J. Lipton (✉)
Abt Associates, 1881 Ninth St., Suite 201, Boulder, CO 80304, USA
e-mail: Josh_Lipton@abtassoc.com

K. LeJeune
Le Moulin du Siffré, 66730 Rabouillet, France
e-mail: klejeune@icloud.com

J. Lipton et al. (eds.), *Equivalency Methods for Environmental Liability*,
https://doi.org/10.1007/978-90-481-9812-2_4

- Quantify damage (by comparing the level of resource and service quality post-incident to baseline conditions); and
- Calculate the interim loss and total debits.

This chapter describes elements of damage determination and quantification that should be considered in an assessment.

Box 4.1: Key Issues and Actions in Determining and Quantifying Environmental Damage

The purpose of damage determination and quantification is to quantify the damaged resources or services that should be offset by remediation projects. Key elements of damage determination and quantification that should be considered in an assessment include the following:

During the *identification of damaged resources, habitats, and services* phase, data and information are analysed to produce a logical and credible estimate of the types of resources or habitats damaged or the services normally provided by the resources or habitats. The damage determination may address not just ecological harm and associated service losses, but social and economic factors that depend on the ecological integrity of the resources in question.

During the *determination of the causes of the damage* phase, the analyst must define, to the extent practicable, the causal linkage between the incident

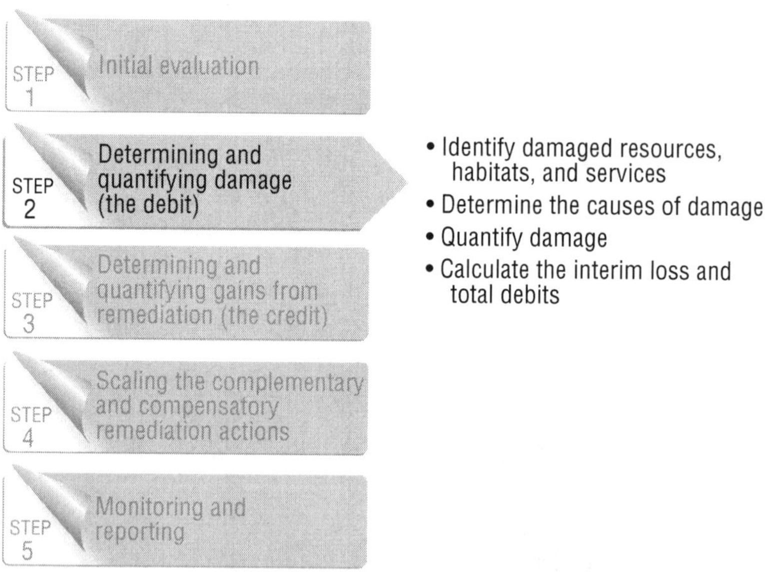

Fig. 4.1 Step 2 of equivalency analysis

and the resulting damage. Scientific data, logical analysis, modelling, and deductive reasoning can all contribute to the characterisation of causality.

During the *quantification of the damage*, the spatial and temporal extent and the degree of the natural resource damage or service loss are quantified relative to baseline conditions. Approaches to estimating the extent and degree of natural resource damage or service loss can include the use of chemical, toxicological, biological, or economic data; geographic information systems; and modelling.

The expression of the degree of damage to the resource or to the services provided by the resource involves two important aspects: the expression is done relative to *baseline conditions* and is typically in terms or one of more measures called *quantification metrics* that are used to quantify the adverse effects of the incident and scale the remediation.

Metrics range from easily measured, quantitative attributes (e.g., population density or user visits), to more conceptual or qualitative attributes and in some cases to complex indexes. The same metric must be used for estimating both the losses and gains.

Baseline conditions are the conditions of the resource or habitat without the incident in question. Baseline conditions can be determined using data collected at the site before the incident occurred, from data collected from reference sites, i.e. sites that are sufficiently similar to the damaged site, unaffected by the incident, or through the use of models. In other instances, damages may be quantified as an incremental loss without explicit determination of a baseline condition.

Once the damage that has been caused or is expected to be caused has been quantified, the next step is the calculation of interim losses and the total debits. Interim losses are calculated by estimating the degree of natural resource or service loss each year between the time the damage occurs and the time that the resources and services are restored to baseline conditions. The degree of loss each year of the damaged area is summed (and discounted) to give the total debit. Calculating interim losses and total debits requires knowledge about the benefits and potential collateral damage of primary remediation and associated (natural and remediated) recovery rates.

4.2 Identify Damaged Resources, Habitats, and Services

Data and information gathered during the preliminary evaluation should facilitate the identification of potentially damaged resources, habitats, and services. These data, plus additional data that may be collected, are analysed to produce a logical and credible estimate of the types of resources or habitats damaged and the services normally provided by those resources or habitats.

Potential pathways of transport or cascading harms,[1] including groundwater and surface water pathways; soil, sediment, and pore-water pathways; food chains and other biological pathways; and aerial pathways should be considered when identifying damaged resources and habitats. Conceptual site models (CSMs) may be used to construct a coherent picture of habitats and ecosystems to help identify potentially affected resources (see Sect. 4.2.4).

Because the equivalency analysis might be focused on the loss of service rather than the loss of the resource itself, it often is important to identify the services normally provided by the impaired resource or habitat. The damage determination should address not just ecological harm and associated service losses, but social and economic factors (use and non-use values) that depend on the ecological integrity of the resources in question.

Although each incident will have different characteristics, the types of data that are typically evaluated when identifying damaged resources, habitats, and services include:

- Site hydrology, geology, ecology, biogeochemistry, and special resources. Information on the presence of European protected habitats and species as defined by the Environmental Liability Directive (ELD) should be included. Habitats are listed in Annex I of the Habitats Directive (HD), species are listed in Annex II and IV of the HD, and wild bird species are referred to in Article 4 (2) or listed in Annex I of the Wild Birds Directive (WBD). It may be necessary to gather information on nationally protected habitats and species if they are referred to in any national legislation implementing the ELD.
- Type of water body. It will be important to obtain information on the type of water body affected, as defined by Article 4 of the Water Framework Directive (WFD). Water body characteristics, a review of the impact of human activity on the status of surface and ground waters, and an economic analysis of water use in each river basin district will be available from each Member State in accordance with Article 5 of the WFD. These data should be collated as part of the preliminary evaluation.
- Site designation, if any. Reference should be made to international site designations (Special Protection Area (SPA) under the WBD, Special Area of Conservation (SAC) or Site of Conservation Importance (SCI) under the HD, Wetland of International Importance under the Ramsar Convention). Reference should also be made to national nature conservation sites if they are included in national legislation implementing the ELD.
- Nature of the chemical contaminants released and their behavior in the environment or nature of the physical stressor when the incident is related to physical disruption of the environment.

[1]The term *cascading harms* refers to indirect or secondary effects that may occur as a result of an incident. For example, a chemical spill may affect primary producers or prey species directly. Secondary producers or predators that rely on these species, but are not directly exposed to the spilled chemical, might then be damaged through indirect, food-chain responses.

- Chemical concentrations in soil, surface water, groundwater, biota, and air.
- Background concentrations of the contaminants of concern.
- Transport and exposure pathways, including cascading harms.
- Physical features of the ecosystem and their vulnerability to the incident. This should include information on the area or extent of habitats impacted and the numbers of animals or, in some cases, plants killed or injured. Air photography and other remote-sensing methods may be used to capture information concerning the extent and magnitude of an incident.
- Potentially affected species, habitat uses (e.g., uses of the site for migration, spawning, rearing, foraging), important trophic relationships, and community composition.
- Important habitat features, uses, and condition.
- Geographic location relative to population centers.
- Recreational and other uses of resources in the area.

4.2.1 Describe the Nature of the Stressor

Characterisation of the damage should take into account the spatial and temporal extent of the harm, as well as the degree of the harm. This step should include a thorough evaluation of the nature of the stressor in question.

For example, for incidents involving the release of chemical contaminants, the following factors generally should be evaluated: the contaminant's chemical structure, its behaviour in the environment, its toxicity, and environmental interactions that may occur to alter its behaviour and toxicity. The evaluation should also consider transport, degradation, uptake, and accumulation pathways of parent compounds and metabolites, as well as secondary releases that may occur as a result of chemical interactions in the environment. Additionally, any additional substances used for primary remediation (such as oil dispersants) should be documented and their effects on the behaviour of the primary contaminants in the environment evaluated.

Biological stressors may include introduced biological agents (such as genetically modified organisms (GMOs); Box 4.2), pathogens, and invasive species. The nature of the biological stressor should be described in terms of the nature of its interactions with baseline ecological communities and species and should include consideration of ecological processes (e.g., nutrient dynamics, decompositional processes), community composition, biological or genetic diversity, predator-prey dynamics, and other relevant considerations.

Box 4.2: The ELD and GMOs

The release of GMOs into the environment may pose substantially different risks to the environment than many of the other activities covered by the ELD

(GeneWatch UK 2005). There currently is relatively little experience or expertise to draw upon in long-term assessment of the potential effects of GMOs. Consequently, potential adverse effects may be less certain and less predictable than for other, more 'traditional' environmental stressors. Finally, the range of potential effects of GMOs might go beyond the protections provided by the ELD. For example, release of GMOs might have an impact on biodiversity as a whole (not just protected habitats and species) or cause damage to land without posing significant human health risks. It should be noted that the ELD approach toward GMO-risks differs from other European Union laws and regulations dealing with GMOs, such as the Deliberate Release Directive (DRD—2001/18/EC), the main European Union legislation governing the use of GMOs in Europe. Under Article 12 of the DRD, an environmental risk assessment must be carried out, including an evaluation of the direct, indirect, immediate and delayed effects that may arise from the release of a GMO. The scope of this evaluation is very wide and not restricted to specified habitats and species as is the case in the ELD.

The ELD takes a general approach and does not treat 'GMO damage' as a separate category. Instead, the ELD provides a strict environmental liability regime and allows Member States to introduce specific levels of protection if they wish to, amongst other reasons, better align the liability regime for GMOs towards existing laws and regulations on their use. The national GMO liability regimes are important as the assessment of damages from GMOs (and hence the level of compensation) should be based on the requirements they contain.

The description should consider potential harm at various scales of ecological organisation (e.g., individual, population, community, and ecosystem) and at various scales of physiological organisation (e.g., subcellular, cellular, organism, and population) (Box 4.3).

Other potential stressors may be physical in nature. Physical stressors can lead to direct loss of habitat or organisms; influence hydrologic regimes or land cover types; and affect water quantity, velocity, important seasonal water level fluctuations, maximum temperature, and erosive potential. A hydrological stressor might be described in terms of too much, too little, or poorly timed water availability or in terms of a change in connectivity to essential hydrologic pathways. Incidents that involve major changes in vegetation cover (such as construction projects or forest fires) might also be described in terms of the resulting hydrological stress on adjacent ecosystems. Because physical changes might have cascading effects, they should be examined closely and described to ensure they are considered in the quantification.

4.2.2 *Exposure Evaluation*

The purpose of the exposure evaluation is to assess the nature, timing, duration, and location of exposures of potentially affected resources or habitats to stressors associated with the incident. The severity of the exposure and any possible secondary exposures should also be assessed.

4.2.2.1 Nature, Timing, Duration, and Location

Factors to consider when evaluating exposure include the nature of environmental exposures; the timing of any exposures (e.g., continuous versus intermittent; relationship to other environmental factors such as daily shifts in dissolved oxygen concentrations, hydrological factors, local tides; relationship to biological factors such as migratory behaviours and spawning cycles); the duration of exposures (e.g., acute versus chronic; continuous versus intermittent; multi-generational); and the location of exposures (including consideration of spatial uses of different habitats by potential receptors; localised physical, hydrological, biogeochemical, and ecological factors that may influence exposure).

4.2.2.2 Severity

When characterising the degree, or severity, of an exposure, contaminant concentrations (in the case of a chemical incident), the degree of physical alteration (in the case of a physical stressor), and the extent of biological exposure to a biological stressor should be considered. In addition, localised factors that could increase or decrease the severity of exposure should also be considered. Factors that might affect the severity of exposure could include:

- Site-specific geochemical conditions;
- The medium into which the contaminant is released and spread (pathway);
- Climatic and hydrologic variables;
- The location of a release or physical impact relative to potential escape pathways or refuges for biological receptors;
- Seasonal factors that might affect the number, age, sensitivity, resistance, or resilience of an exposed resource or habitat; and
- The potential for delayed, cascading, or synergistic adverse effects.

4.2.2.3 Secondary Exposures

Secondary exposures are possible if a released substance reacts in the environment and produces a by-product. This can be a concern for some environmental contaminants. For chemicals that bioaccumulate or bioconcentrate, secondary exposure may occur through food-web interactions.

Secondary physical and biological effects are also possible. For example, if a chemical release eliminates vegetation that stabilises stream banks or moderates temperature, fish might be exposed to excessive sedimentation or temperature as a secondary effect of the release. Potential effects or 'risk cascades' (Lipton et al. 1993) such as these may be relevant to exposure evaluations.

4.2.3 Receptor Evaluation

Receptors are the organisms, communities, habitats, ecosystems, and services that are exposed to the effects of the incident. Depending on the incident and its effects, receptors can be described at several scales of organisation, ranging from suborganismal (cellular) to ecosystem-level endpoints. Numerous receptors can be used in, or at least considered for use in, an equivalency analysis. Receptors can include:

- Suborganismal endpoints, such as biochemical receptors, target tissues, biological or cellular processes, and genetic receptors;
- Individual biota, such as fish, benthic macroinvertebrates, birds, and mammals;
- Populations of biota;
- Biological communities, such as cold-water streams, emergent wetlands, mudflats, and riparian forests, if an incident has broader effects on ecological organisation;
- Habitats or habitat assemblages;
- Landscapes; and
- Ecosystems or ecosystem processes.

It is important to note that specific receptors, levels of organisation, and endpoints may be specified in different European Union environmental Directives and laws of Member States.

Receptors can also include human endpoints. Humans may be directly exposed to chemical releases, indirectly exposed through food-chain transfer, or secondarily affected by restrictions, bans, shortages, or similar effects of resource or services losses.

4.2.3.1 Secondary Receptors and Damage Cascades

Some effects of the incident may only be revealed or discovered with time, as chemical contaminants reach receptors and as the environment readjusts to a new equilibrium following the disturbance. Sometimes, the cascading effects of an incident are subtle, delayed, or are experienced in a location removed from the primary site of the incident (Lipton et al. 1993). For example, if chemical contamination reduces the vigour of vegetation and soil biota, and severe erosion occurs as a result, the cascade of effects can denude a habitat or a structure that formerly provided shelter, forage, or other essential needs of wildlife and human use services such as shoreline buffering or flood storage.

In evaluating damage, analysts should think broadly about potential secondary and cascading effects. Damage at one spatial scale can have a cascading effect on a very different type of service at another scale. For example, a primary effect of an incident might be to eliminate primary productivity in a stream. If that causes adverse effects on a downstream fishery which is economically important for a human population, the landscape-level cascading effect could be to cause economic hardship for the affected population or increased fishing pressure and related adverse effects in a neighbouring water body.

4.2.3.2 Levels of Organisation

Receptor evaluation should take into account harm at various scales of ecological organisation (e.g., individual, population, community, ecosystem) and at various scales of physiological organisation (e.g., subcellular, cellular, organism, population). Sometimes the receptor at a landscape level is different from the receptor at a smaller scale. At the landscape level, the receptor might be a human population (and the effect might be a social or economic loss), whereas the receptor at a smaller scale might be fish or a wetland (and the effect might be death or destruction).

Box 4.3 presents background information related to defining affected populations and levels of organisation.

Box 4.3: Defining affected populations and levels of organisation

In contrast to human health risk assessment, some ecological assessments may focus on establishing risks/damages at the population level. In broad terms, a population in biology is a group of interbreeding organisms occupying a particular space or the number of humans or other living creatures in 'a designated area'. Alternatively, a population is sometimes defined as a group of organisms of the same species 'relatively isolated' from others of the same species. Both definitions leave considerable margin to delimit the boundaries of the area or interpret what is to be considered 'relatively isolated'. This vagueness is transferred to the concept of 'affected population' that can be regarded as that population that suffers effects from an event or activity.

In United States environmental liability cases, defining affected populations has resulted in ongoing tension. When it is too broadly defined, it may be difficult to determine damage. If the affected population is too narrowly defined, it becomes easier to establish that there is damage but it may be harder to articulate a case for broad societal value. For example, an incident can cause catastrophic impacts to a local population without discernibly influencing an 'overall' population (e.g., within an entire flyway[2]).

In the European Union, different regulations and legislation refer to different levels of biological organisation. Some stipulations make it possible to take into account damages to individual organisms, while several elements favour assessments at higher level of organisation. In addition, national legislations might slightly differ in this respect from the European Union level and other countries. It is therefore not possible to define a generic level of organisation at which 'biodiversity damage' should be established. It is the responsibility of the Competent Authority to make a judgment on the appropriate level of organisation to be assessed based on the best available scientific information and practices while taking into account the relevant legal framework (European Union and national/regional) and the specifics of the damage and affected site.

However, even when looking at higher levels of organisation, it is often useful and sometimes even necessary to assess effects at lower levels of organisation because damages at those scales are much easier to establish and can often serve as a proxy for damages at higher levels of biological organisation. Hence, it is recommended to start establishing effects at a lower level (e.g., at the individual level) and then work up to higher levels (populations, communities, ecosystems). Depending on the situation, effects on individuals/populations/communities might be more or less relevant (e.g., protected status, rarity, importance for ecosystem integrity and services, and so on). This requires determining the importance of the observed effects in a larger ecological and sometimes socio-economic framework.

[2]The terms 'migration route' and 'flyway' are to some extent theoretical concepts. Migration routes may be defined as the lanes of individual travel from any particular breeding ground to the non-breeding quarters of the species (e.g. birds, fish) that use them. Flyways, on the other hand, may well be conceived as those broader areas in which related bird migration routes are associated or blended in a definite geographic region. They are wide arterial highways to which the routes are tributary. Except along the coasts, flyway boundaries are not always sharply defined.

4.2.4 Conceptual Model of Exposed and Affected Resources and Habitats

As receptors are evaluated and as impacted organisms, communities, ecosystems, and services are identified, a conceptual model of the potentially damaged system should be developed. This step can be a simplifying stage, where the main adverse effects and indicator resources and habitats are selected for inclusion in an equivalency analysis. This step is often necessary because ecosystems are complex and the parts interrelated. If all the damage done is quantified and added independently, the amount of remediation indicated might overstate the true amount needed, since a single remediation project might address multiple resource damages.

A CSM is a basic description of how contaminants enter a system, how they are transported around within the system, and where routes of exposure to organisms and humans occur. As such, it provides an essential framework for assessing risks from contaminants, developing remedial strategies, determining source control requirements, and addressing unacceptable risks. CSM development typically entails identifying relevant habitat types, identifying likely exposure pathways and linkages between abiotic and biotic receptors, identifying primary tropic categories and representative species, and identifying important ecosystem functions or processes.

4.2.5 Damage Determination

After characterising the stressors, receptors, and the pathway linkages between them, the next step in the equivalency analysis is to determine what damage to natural resources has occurred. Damage determination is broadly defined as the demonstration of an adverse change in the biological, chemical, or physical quality of a natural resource or service. This broad definition allows Competent Authorities and analysts some freedom in defining the characteristics of the damage. However, it does not relieve the burden of presenting a logical assessment of the adverse change and its relationship to loss of services provided by the resource.

Although the Toolkit presented in this book does not provide insight into whether damage is to be considered ‘significant’ or not (and this may vary across Member States, depending on individual legislation), some elements and steps of the methods presented here may be of use when establishing significance. It could be argued that any measurable adverse effects are significant. Therefore, the determination should include evaluation of the baseline state of the resources to aid in determining significance as well as the degree of harm caused by the incident.

4.2.6 *Types of Impacts to Natural Resources and Their Services*

The following section identifies common types of damages and service losses associated with classes of natural resources. Although this listing is not comprehensive, it can be used to guide thinking about specific incidents and effects.

4.2.6.1 Surface Water

Surface water can be damaged when chemical, hydrological, or physical conditions of a surface water body are sufficient to cause adverse effects to aquatic biota or to human users of the water. Damages to surface water may include exceedances of water quality standards (including those for biota, drinking water, recreational uses, and agricultural uses; toxicological or flow-related thresholds; criteria for protection of aquatic biota; other numerical or verbal criteria intended to protect humans and other biota).

Surface water can be damaged when surface water conditions are such that the water causes lethal, sublethal, behavioural, genetic, immunological, or other adverse effects in individual aquatic organisms, populations, communities, or ecosystems. Damages to surface water may also include changes in the water that are not necessarily regulated or even the primary result of the incident but that cause harm as a secondary exposure or risk cascade. For example, excessive turbidity, insufficient dissolved oxygen, temperature swings, pH changes, or resuspension of buried chemicals might occur as a secondary effect of an incident or the response to an incident. In addition to chemical changes in surface water, hydrological, physical, or other flow-related alterations also could constitute damage.

In assessing damages to surface water, factors that might influence the effect of a primary causal agent should be considered and measured. For example, constituents that might influence toxicity of metal contaminants released to surface water might include pH, temperature, calcium, magnesium, iron, and dissolved organic carbon concentrations.

Damages to surface water can also include closures of public access to surface water bodies or restrictions on public uses (such as fishing and swimming), even if the response to the incident, rather than the incident itself, prompted the closure or restriction.

Examples of service losses associated with surface water impacts might include disruption of drinking water supplies, recreational use (swimming, boating, fishing) closures for some duration of time, adverse effects on aquatic biota or habitat, and accumulation of substances in the food chain that lead to adverse effects on biota. Additional service losses that might be considered include reduction in the assimilative capacity of the water (the ability of surface water to absorb low levels of contaminants without exceeding standards or without adverse effects), hydrological alterations (including quantity or timing of flow), and the 'stigma' associated with contamination. Stigma captures the idea that even if cleanup actions might be complete, people prefer not to use or visit the site because of the occurrence of the incident, or that clean-ups may not be completely effective or certain and, therefore, there is a continued perceived risk (and associated loss of economic value).

4.2.6.2 Groundwater

Groundwater can be damaged when chemical, hydrological, or physical conditions in an aquifer are sufficient to adversely affect human uses of the water, or biota or habitats exposed to a discharge of the groundwater. Damages to groundwater might include exceedances of drinking water standards or guidelines; toxicological or hydrological thresholds or criteria for biota that might be exposed at seeps, springs, or gaining sections of a river or bay; or other numerical or verbal criteria for groundwater intended to protect humans and other biota.

Physical or hydrological damages might include reductions in the aquifer's water-holding capacity, reductions in the safe yield from an aquifer, alterations of recharge/discharge relationships, or destruction of an aquifer by compaction or sealing in such a way that a source of groundwater or groundwater-dependent habitat is no longer available. Physical restriction of access that results when an aquifer is being used for other purposes might also damage groundwater.

Service losses associated with groundwater impacts might include disruption of drinking water supplies for humans or livestock, preclusion of future use of an aquifer as a public drinking water supply or for use in agriculture, closure of a recreational use area because of the risk associated with a groundwater plume, or habitat degradation related to the toxicity of substances in shallow groundwater. Physical damage to an aquifer might cause, directly or indirectly, similar disruptions of services.

4.2.6.3 Sediments

Damage to sediments can be assessed using relevant guidelines and standards or by demonstrating that the incident has affected sediments such that they adversely affect other resources. Various agencies and organisations in Europe and North America have developed numerical sediment quality guidelines, and sediment toxicity tests using a variety of approaches have been performed to assess the quality of freshwater and marine sediments (Long et al. 1998; NOAA 1999; CCME 1999; Macdonald et al. 2000; Crane 2003; Babut et al. 2005; Hin et al. 2010). The approaches selected by individual jurisdictions differ based on the ecological receptors considered, the degree of protection afforded, the geographic area to which the values are intended to apply, and the intended uses of the values. In Europe, the WFD calls for the development of sediment Environmental Quality Standards and proposals have been made for achieving this. However, these proposals are technically controversial and could have considerable logistical consequences (Crane 2003).

Actual measurements can be compared to guidelines or thresholds in order to demonstrate the probability that measured chemical concentrations are (or will be) causing adverse effects on biota exposed to the sediments. For example, MacDonald et al. (2000) assembled previously published sediment quality guidelines for 28 chemical substances and classified them into two categories: a Threshold Effect Concentration (TEC) and a Probable Effect Concentration (PEC). TECs are intended to identify contaminant concentrations below which harmful effects on sediment-dwelling organisms are not expected to occur. TECs include threshold effect levels; effect range low values, lowest effect levels, minimal effect thresholds, and sediment quality advisory levels. PECs are intended to identify contaminant concentrations above which harmful effects on sediment-dwelling organisms are expected to occur frequently. PECs include probable effect levels, effect range median values, severe effect levels, and toxic effect thresholds. The published sediment quality guidelines were then used to develop consensus-based TECs and PECs (MacDonald et al. 2000). Analysts can use these and similar thresholds to evaluate damage to sediments.

Physical damage to sediments might include scouring, burial, or alterations in grain size distribution. Such effects can adversely affect the ability of sediments to provide habitat for sediment-dwelling and sediment-feeding organisms. These types of damages might be described as damage to sediment, habitat, or biota.

Service losses associated with damage to sediment might include elimination or reduction of the ability of the sediment to provide habitat for aquatic biota, including plants, benthic invertebrates, fish, and sediment-feeding birds. Damage to sediment might cause wetland plant kills, reductions in wetland or aquatic plant cover, shifts in community composition, or simplification of wetland or aquatic plant community structure, which diminishes wildlife habitat quality. Human use service losses might include reduced access to recreational areas or reduced quality of experience at a recreational area.

4.2.6.4 Soils

Examples of soil damage in protected habitats could include concentrations of chemicals that cause toxicological responses to soil microorganisms, invertebrates, plants, or wildlife (e.g., Efroymson et al. 1997a, b). Physical damages to soil might include erosion or burial, alterations to soil structure or function (e.g., water-holding capacity, nutrient cycling), or loss of supporting habitat for biota.

Service losses associated with damage to soils might include elimination or reduction of a soil's ability to provide habitat for wildlife or grazing for livestock. Injuries to soils might cause plant kills, reductions in plant cover, shifts in community composition, simplification of plant community structure that diminishes wildlife habitat quality, or grazing land quality. Human use service losses might include reduced access to recreational areas, reduced quality of experience at a recreational area, or reduced access to grazing, or resource extraction on public lands.

In addition to ecological effects of contaminated soils, soil damages may be confirmed through a determination that the damage poses a risk to human health. Although national laws transposing ELD may use stricter definitions, the ELD covers soil damage only if it poses a human health risk. However, there are two situations in which resource equivalency might be needed when soil damage has occurred or will occur: (1) when the damage is not limited to soil/land but also results in damage to protected habitats, species, or water; and (2) when primary remediation measures to remove the health risk cause damage to protected water, habitat, or species covered by the ELD are not sufficient.

4.2.6.5 Vegetation

Damage to vegetation can include reductions in cover, diversity, health, vigour, reproductive capacity, stability, or habitat value of plants. In addition, reduction of the nutritional or habitat value of plant species for wildlife species of concern may also constitute damage.

Damage to vegetation is commonly a result of physical incidents. However, in such cases, it can be relatively straightforward to identify and describe it. More subtle and difficult to identify are incidents of damage to soil that cause a change in the soil's biogeochemistry or the plant composition of the remaining community, such that it is no longer possible to re-establish baseline vegetation. For example, physical disturbance may allow invasive, undesirable plant species to become established. These species then out-compete native plant species, causing a permanent and detrimental shift in the community composition.

Service losses associated with damage to vegetation include degradation of wildlife habitat quality, degradation of the physical stabilisation that vegetation provides to soils, and reduction in thermal or hydrological attenuation attributes of vegetation cover. Loss of, or damage, to vegetation can cause reductions in recreational use opportunities or enjoyment, losses of food or medicinal plants, loss of landscape value, or reductions in non-use values.

4.2.6.6 Biota

Damage to biota can take many forms. Damage endpoints could include death of individual organisms, population reductions, changes in community composition, loss of supporting habitats, and lethal effects that influence the viability of organisms or populations. Sub-lethal effects might include disease (or compromised immune systems), behavioural abnormalities, cancer, genetic mutations, physiological malfunctions (including malfunctions in reproduction), or physical deformations. In general, the result of any stressor that causes the biological resource to undergo adverse changes in viability might be considered as damage. In addition, instances when a stressor, such as a chemical contaminant, causes concentrations in edible tissues of biota to exceed safe consumption levels, might be considered damage.

For aquatic biota, exceedences of surface water or sediment criteria could indicate damage to fish and benthic invertebrates. Assessment could be supplemented with an evaluation of toxicological thresholds derived from the literature. In developing toxicological thresholds, analysts should consider test species and their relative sensitivity to contaminants and site-specific water quality conditions that may influence toxicity. Depending on the circumstances, analysts might consider conducting site-specific toxicity tests to evaluate potential acute and chronic effects of contaminants.

Population data may be used to determine if spatial patterns of organism abundance, diversity, or age structure are indicative of damage. Fish or wildlife kill investigations, necropsy, pathology, and chemical analyses, information on organism reproduction, and available literature might all provide information that could be used to determine damage. Both field and laboratory investigations can be used to determine damage.

4.2.6.7 Habitats

Environmental damages also can be manifested at the habitat level. To assess damage to habitats, analysts can compare key habitat attributes, such as (for terrestrial habitat) vegetation cover, composition, structure, forage quality and production, or thermal cover in assessment and reference sites. Loss of vegetation components can adversely affect wildlife that depends on the vegetation for forage, nesting, staging, hiding or thermal cover. Examples of key attributes for aquatic habitats include: physical habitat structure, river channel characteristics, sediment/substrate quality or quantity, characteristic community structure and species composition (including benthic communities/species), flow regimes, trophic interactions, water temperature, nutrient regime, light penetration, water quality, and sediment regime.

Habitats can also be impacted indirectly by release of contaminants, biological agents, or physical damage. If an incident causes changes in nutrient cycling that reduce productivity or community composition, the services provided by the baseline community may be diminished. For example, loss of vegetation or changes in types of vegetation can lead to erosion of soils and stream banks and subsequent effects on biological communities that depend on clean, sediment-free water or stable stream banks. Analysts should consider such cascading or downstream effects on all affected habitats, including the indirect physical effects that result from the impacts of an incident.

To simplify an assessment, sometimes an indicator species in a habitat can be identified if ecological associations can be scientifically established.

4.2.6.8 Human Use and Non-use Values

Natural resources provide many services to humans, currently referred to as ecosystem services (see Chap. 2, Box 2.3). Loss of human services can result from the damage, or threat of damage, to natural resources and can result from changes in either the quantity or quality of the services provided by resources. Service losses can be associated with actual or potential risks[3] to human health; with lost uses (or potential uses) of natural resources; or with losses of non-use values provided by natural resources.

Loss of human services can be measured directly and/or indirectly. Direct measurements include the change in the quantity of resource used, such as the number of fishing or hunting trips taken by individuals or measurable increases in the risk to human health as developed through epidemiological studies. Monetary values for these services can be estimated using prices and consumer behaviour from actual markets. Other services do not have actual markets in which they are traded. Yet others may not even be used directly (e.g. regulatory services). Different valuation methods are developed to collect and analyse data when markets are not missing (non-market values) and when services are not used directly (see Chap. 8 for further details).

4.3 Determining Causes of Damage

Multiple stressors can influence natural resources and their services. Some stressors are natural and can be relatively constant, periodic, or episodic. Other stressors are anthropogenic but unrelated to a specific incident in question. The analyst should identify the effects of the incident in question in order to determine and quantify the damages associated with the incident. The analyst also should define as precisely as practicable the causal linkage between the incident and the resulting damage.

Scientific data from the literature, logical analysis, site-specific studies, modelling, and deductive reasoning can all be used to evaluate causality. Determination of a likely or probable causal link between the incident and the change in resource condition or services provided can be subject to intense scrutiny. Thus, the science or economics involved in establishing causality should be rigorous and transparent.

Determination of the cause of a given damage might involve field or laboratory investigation or original research if a chemical, biological, or physical effect is complex, rare, or relatively poorly understood. In many cases, a single agent may not be the sole cause of the environmental damage in question. In determining

[3]Risks to human health can result from direct physical contact with contaminated resources (e.g., soil, water), ingestion of contaminated land or food sources (e.g., soil, plants, fish, meat), or inhalation of contaminants. Risks can be associated with both lethal and sub-lethal effects such as reduced reproductive capacity, reduced mental capacity, or increased respiratory disease. The ELD defines any damage with a proven effect on human health as significant.

causality, it may not be necessary to determine the precise effect of the single agent on the natural resources and services involved. Simply demonstrating that a causal link is plausible and likely to have been at least a partial contributor to the effect might be sufficient.

4.4 Quantifying Damage

After the nature of damages has been determined, those damages should be quantified. Quantification of damage typically includes an evaluation of:

- The spatial extent of damage and service or resource loss;
- The temporal (past, present, and expected future) extent of damage and service loss; and
- The degree of damage and loss of habitat, resources, or services (often expressed as a percent of services provided relative to baseline conditions, in terms of numbers of organisms, or as a reduction in the quality of a characteristic of the organism or habitat).

The extent and degree of damage and service loss can be estimated using chemical, toxicological, biological, or economic data; geographic information systems; and modelling.

Characterisation of the spatial extent of damage should entail identifying the full areal extent of damages and might include identification of damage gradients or impact zones. Sampling or modelling to assess transport, dispersion, dilution, transformation, or adverse effects might assist in identifying the damage gradients or zones. Photographs and remotely sensed data (e.g., aerial photographs, satellite imagery) can be used in some instances to identify the spatial extent of impacts, and nested wells or core samples can be used to identify the depth of contamination in an aquifer. Sampling data, remotely sensed data, and spatial analytical methods can be used together to model the spatial extent of damage.

Characterisation of the temporal extent of damage involves identification of the date of the incident and the date when the adverse effect occurred (if the two differ). For imminent threat, the likely date of damage could be estimated based on the chemical or physical damages expected. For known future events (e.g. planned projects), the start date might be a projection based on stated goals, plans, or schedules. If site-specific information is not available to quantify the temporal extent of damage, the duration of losses may be based on knowledge of similar events at similar sites. Recovery trajectories might be estimated based on ecological succession rates, chemical persistence in the environment and understanding of fate and transport dynamics, or information from published literature on recovery rates following similar disturbances. If primary remediation actions are planned or underway, the estimate of the temporal extent of the damage should take into account the anticipated effects of the remediation on recovery.

The degree of damage or service loss generally should be expressed relative to baseline conditions. In some instances, this will be performed by explicitly quantifying baseline and post-incident conditions. In other cases, only the distinct, or differential damage caused by the incident will need to be calculated (e.g., by calculating the differential mortality caused by a chemical toxicant or by quantifying the amount of physical damage to a habitat type from a development project). The degree of damage to the resource or to the services provided by the resource is typically expressed in terms of one or more measures that can be used to reflect the adverse effects of the incident. These measures are called metrics. Selection of metrics and characterisation of baseline conditions are discussed in more detail in the sections that follow.

4.4.1 Quantification Metrics

Quantification metrics are used to express both the degree of natural resource damage or service loss attributable to an incident and the degree of benefits attributable to a remediation project. The selection of an appropriate metric is important because the amount of estimated loss and gain can vary depending on the metric used.

Metrics can include both readily measured, quantitative attributes, such as population density, vegetation cover, productivity estimates, or user visits, as well as more complex, conceptual, and qualitative attributes, such as habitat suitability or quality indices, multi-variate indices, or subjective rankings.

The metric should be the same on the loss (debit) and gain (credit) sides of the equation to enable equivalency calculations. If the metrics are different, it may not be possible to balance the debits and credits without adjustment factors or scalars, which is the objective of a resource equivalency analysis. For example, if the degree of loss is quantified in terms of percentage of native vegetation cover per hectare remaining after an incident, the gain from restoration should be also in terms of percentage of native vegetation cover per hectare created. If the loss is quantified as a population reduction relative to baseline populations, the gain should be measurable as a population increase relative to baseline. Examples of quantification metrics include:

- Extent of a specific habitat type;
- Units, or quanta, of some resource (e.g., kilometers of a type of river, hectares of a specific habitat type, volume of usable water);
- Measures of vegetation density, cover, or biomass;
- Percent cover of a desirable, dominant, or essential vegetation species;
- Aboveground biomass of the dominant vegetation;
- Density of seedlings;
- An index of vegetation structural diversity;
- Habitat-quality indices;

- Biological productivity (e.g., primary or secondary productivity), species abundance, biomass, diversity, or measures of community composition;
- Reproductive rates;
- Habitat use-days (e.g., when an incident has reduced the availability of habitat such that fewer organisms can occupy the habitat);
- Indices of population integrity such as sex ratios, age class distribution, biomass;
- Measures of ecological processes, such as rates of carbon mineralisation, nutrient export, or decomposition;
- Categories of service loss assigned based on the degree of exceedance of toxicity thresholds (e.g., this approach might involve compiling dose-response information from the literature or site-specific studies and developing an estimate of service loss as a function of increasing contaminant concentration in soil, sediment, surface water, or biological tissues.); and
- For a value equivalency analysis (including both value-to-cost and value-to-value), money can be regarded as the metric to measure the damage, and remediation (for value-to-value).

Damage quantification can be conducted on a resource-to-resource basis; in many cases, this is a logical way to proceed. However, natural resources and the ecological services they provide are interdependent. For example, surface water, sediments, floodplain soils, and riparian vegetation together provide both habitat and lateral and longitudinal connectivity between habitats for aquatic biota, semi-aquatic biota, and upland biota. Therefore, damage to individual natural resources may cause ecosystem-level service reductions. Analysts should consider these interdependent ecosystem-level service losses when selecting metrics and quantifying service losses.

Multiple measures of service provision include published or accepted indices for the health of the environment, as well as indices developed for specific incidents and habitat equivalency analysis applications. A useful text describing the uses and potential misuses of multi-attribute indices can be found in Ott (1978).[4] Examples of these types of measures for use in equivalency analysis are described in the following paragraphs.

Indicators of habitat suitability are commonly used to aggregate numerous attributes related to thermal and hiding cover, forage availability, reproductive requirements, and the capacity of the physical habitat to support characteristic functional and structural communities important for a wildlife species or community of interest. The aggregate score indicates the current and anticipated future condition of the land to provide habitat services.

[4]Using a compound metric necessarily implies a weight to the importance of each of the individual components (e.g., uniqueness and evenness). Because a metric should be the same on the debit and credit side of an equivalency analysis, special consideration must be taken when that metric is an index value (i.e., dimensionless), because the offsetting remediation project should not only provide an increase in the index (metric) but also maintain the original proportional weights between the individual components (e.g., uniqueness and evenness).

A series of decision rules based on numerous risk indicators, such as degree and frequency of exceedence of toxicity thresholds or aquatic life standards, evidence of population shifts or reductions, and evidence of loss of functional groups or biogeochemical pathways. With a set of tailored decision rules, the degree of service loss is assumed to increase with an increasing number of indicators, or rules, satisfied. Such decision rules are typically developed on a case-by-case basis and are site- and incident-specific.

Sometimes it is necessary to identify multiple possible metrics, because one or more might not be suitable on both the debit and credit sides of the equivalency equation (remember that metrics must be the same on both sides of the equation). In addition, sometimes a single metric will not capture all aspects of loss. If more than one metric is used, the analyst must carefully determine whether the losses estimated by different metrics are independent or additive or if there is some overlap between the metrics on either the loss or gain side.

Equivalency analysis outcome is sensitive to the choice of metric used to quantify lost and replaced resources, habitats, or services. For a case study of habitats equivalency analysis in a salt marsh, Strange et al. (2002) found that different metrics of ecological functions (e.g., above-ground biomass, soil nitrogen, density of fauna) resulted in more than a threefold difference in alternative remediation requirements. Because all habitats and natural resources provide a variety of ecological functions and services, a single metric may never capture all losses. Therefore, the choice of a metric is a very important consideration in adequate remediation scaling. To reduce the chance of disagreements at later stages of equivalency analysis, Competent Authorities and operators, in close consultation with biologists, ecologists, or other relevant environmental scientists and economists, as relevant, should cooperate early on about the selection of an appropriate metric.

4.4.2 *Baseline Determination*

Baseline conditions are the conditions of the resource or habitat that would have been present had the incident in question not occurred. To determine and quantify the type and degree of damage and to scale the required remediation, post-incident conditions are compared to baseline conditions.

Baseline conditions can be quantified using pre-incident data from the damaged site or data from similar sites unaffected by the incident, that is, 'reference' sites or by using models.

4.4.2.1 Use of Before-After Data

In some cases, baseline conditions may be adequately documented using existing pre-incident information. For *ex ante* events [as in the context of the HD, WBD, and

Environmental Impact Assessment Directive (EIAD)], characterisation of baseline conditions should always be a precursor to any development or disturbance. For *ex post* events (as in the context of ELD, even for imminent threat cases), baseline conditions can sometimes be reconstructed using site-specific geological, geochemical, biological, and other databases that may have been compiled for very different uses. Characterisation of baseline conditions might include a description of the chemical, biological, and physical conditions of the site before the incident; the social and economic characteristics of the site; and the role of the site in a larger context—ecosystem or economic—if appropriate.

Site-specific monitoring data, including, for example, data on water chemistry and sediment quality, the biological integrity of streams (including information on benthic invertebrate communities), and similar indices of ecosystem health tracked by regulatory agencies are often quite valuable. Records of anthropogenic sources of stressors other than the incident under investigation may also be useful to understand the baseline context.

4.4.2.2 Use of Reference Sites

If pre-incident information about baseline conditions at the damage site is insufficient, data from reference sites may be used to characterise baseline. Reference locations should be selected by considering those factors that can influence the quality and quantity of natural resources or services provided at a given location. Characteristics might include:

- Ecoregion;
- Location;
- Climate;
- Topography;
- Land uses;
- Human population density;
- Stream size, elevation, orientation, and bordering land uses;
- Bay or estuarine configuration, bathymetry, currents, fringing habitats, and bordering land uses;
- Geology, geochemistry, hydrology, and soil types; or other factors that influence or control the abundance or diversity of organisms, habitats, or biological communities; and
- Important demographic factors (e.g., population size, proximity to population centres, access, scarcity, resource management, regional importance) (when establishing baseline human use and non-use values for value equivalency analysis).

When selecting reference sites for assessment of biodiversity damages, it will be important to select sites that support the same European protected habitat types and, in some instances, specific vegetation types within such habitats. These sites can be identified using EUNIS (European Union Nature Information System) habitat classifications or other national vegetation classifications. Reference sites must be within the same biogeographic region as the damaged site and should be as close to the damage site as possible.

Reference site selection will also be influenced by the availability of existing environmental data. Sites with a long continuity of monitoring data are likely to serve as better reference sites than those from which data need to be gathered in order to establish a reference.

Reference sites are selected such that they match the damaged site to the greatest extent practicable. In some cases, multiple reference sites may be utilised and reference conditions may be described in terms of an acceptable or typical range. The choice of reference site or sites might vary depending on the attribute being characterised. In addition, multiple reference sites might be needed to address numerous components of damage at an assessment site.

4.4.2.3 Use of Models

If reference sites are unavailable, inappropriate, or insufficient or if the baseline condition description that is needed is a condition of an organism, then modelling may be the most appropriate approach to determining baseline conditions. Models might be simple and descriptive or complicated numerical codes. Regardless of the degree of detail or sophistication, the model used should be supported by credible scientific logic.

Use of existing, accepted models can expedite baseline condition analysis; some agencies regularly use models in evaluating environmental conditions. However, analysts should be cautious in using existing models since constructing a site-specific model can be more effective than using a model that was designed for other purposes and is a poor fit for the site-specific circumstances.

4.4.2.4 Use of Incremental Loss Approaches

In many instances, quantifying damages in terms of an incremental loss relative to a remediation target, without explicit quantification of baseline conditions, is sufficient for an equivalency. For example, if a localised wetland is damaged by an incident, it may be sufficient merely to characterise the nature and degree of the wetland damage and conduct remediation activities on a similar wetland. Such an analysis can be undertaken without explicit quantification of baseline conditions (such as faunal diversity or abundance) if it can be assumed that the remediated wetland habitat will provide, generally, the same type and level of functional or structural services as the damaged wetland habitat.

4.5 Calculating Interim Loss and Total Debits

Calculating interim loss involves estimating the degree of resource or service loss each year from the time the damage first occurs and the time that the resources and services are restored to baseline conditions (through primary remediation or natural recovery). The degree of loss each year of the damaged area is summed (and discounted) to give the total (present value) debit.

Figure 4.2 illustrates interim service losses (debits) over time following a hypothetical incident. In this example, service losses have been damaged as a result of an incident. Primary remediation benefits natural resources, causing a relatively rapid improvement in conditions. Following primary remediation, conditions continue to improve at the site, ultimately recovering to baseline. The total interim loss, or debit, is the total area between the baseline and the curve showing the level of resources or services over time.

In cases where the damaged resources or services do not recover to baseline, interim losses are aggregated over perpetuity.

4.5.1 Calculating the Total Debit

Below, we describe the conceptual approaches for estimating total debits (loses) for a Habitat Equivalency Analysis (HEA) as an example of a service-to-service approach.

The conceptual approach of HEA to calculating a service equivalency between damage costs and remediation benefits[5] calculates the total debit using the following formula:

$$\sum_{t=0}^{t=n} (A_t \times d_t) \times (1+r)^{(T-t)}$$

where $\sum$ is the summation sign, $t = n$ is the end year, $t = 0$ is the start year, A_t, is the spatial extent (area) of damage (hectare) damaged every year during the relevant period, d_t is the degree of loss (% of service loss) every year, r is the discount (or compound) rate, T is the base year, and t is any given year in the assessment period (between 0 and n).

Below we explain the inputs in this formula and then provide an illustrative example of how this formula is applied to calculate the total debit (loss) measured in Discounted Service Hectare Years (DSHaYs) in Sect. 4.6.

[5]Services here refer to ecosystem services including those that affect human use, i.e. ecosystem services. As noted previously, natural resources themselves often serve as the metric for quantification and the reliance on an integrated measure of services is not universally accepted.

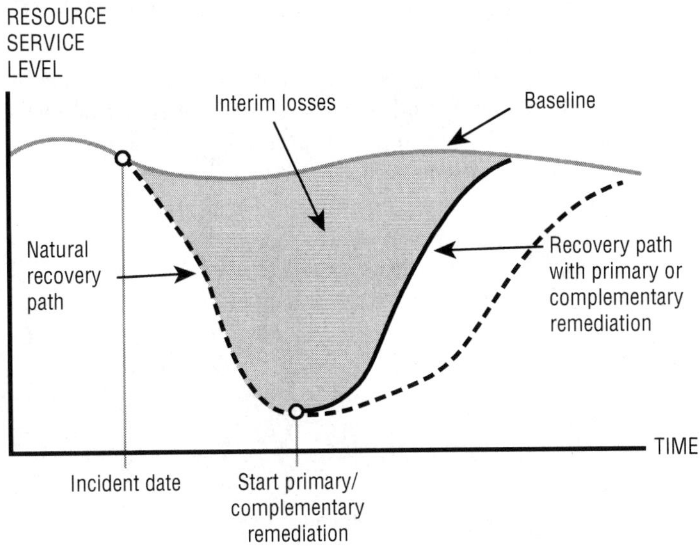

Fig. 4.2 Illustration of cumulative interim losses

- *Start year (t_0).* The start year is the year in which damages began in ELD cases (or are expected to begin in HD, WBD, and EIAD cases or imminent threat in ELD) or the year in which the calculation of losses begins.
- *End year (t_n).* The end year is the year in which no further damage is suffered—either the resources recover naturally or as a result of primary remediation actions. Sometimes there is no expected end year because resources are not expected to recover. In these circumstances of damage continuing into perpetuity, the above formula becomes

$$\sum ((A_t \times d_t)/r)$$

for discounting.
- *Base year (T).* This is the year used for the present value calculations and is generally the year in which the damage assessment takes place.
- *Spatial extent (A_t).* The area of habitat impacted is typically expressed in hectares, but it could be expressed in stream miles or other unit of habitat length, area, or volume.
- *Degree of loss (d_t).* This is the degree of resource or service loss within the spatial extent. It is usually measured as a percentage changed based on expected change in the selected quantification metric. Degree of loss can vary over time.
- *Present value multiplier.* HEA calculations typically incorporate a discount rate of 3% or 4% (see Box 4.4). The discount (compound) rate has the effect of compounding past service loss and discounting future service loss to estimate the present value. Note that since compounding is about bringing past damages

to present value, it is not concerned with the end year. In other words, the end year for compounding is always the base year.

- In a simple case, the degree of loss might be constant over the spatial extent or the loss might decline linearly with time to zero. In a more complicated case, the spatial extent could be divided into damage gradients with different degrees of damage or recovery trajectories.

Box 4.4: Discounting

Debits (losses) and credits (gains) that are delivered in the past or the future are not valued in the same way as those that are delivered today. There are two main reasons for this difference. The first is the individuals' *time preference*, which means, all things equal, individuals prefer to consume now (today), rather than wait. The implication of this impatience is that we need compensation for postponing the consumption of good things (e.g., consumer goods, environmental resources, etc.). The second reason for this difference is *cost of capital*—the resources (money or other) that are available today can be used (invested or processed) to generate further benefits which would be lost if they are not available until later year(s).

Thus, we need to use a procedure to ensure that debits and credits that occur at different points in time are compared on an equal basis. This procedure uses a present value multiplier which takes into account a rate (r) to adjust the future or past values in present terms:

$$(1+r)^{(baseyear-year)}$$

When used to bring past values to present terms, the rate used is called the compound rate and the process is called **compounding**:

$$1 \times (1+r)^{(baseyear-year)}$$

In other words, when the year is less than base year (in the past), the 'power' of the present value multiplier has a positive sign and the present value multiplier becomes a compound factor.

When used to bring future values to present terms, the rate used is called the discount rate and the process is called **discounting**:

$$1 \times (1+r)^{(baseyear-year)}$$

In other words, when a given year is greater than base year (in the future), the 'power' of the present value multiplier has a negative sign, and the present value multiplier becomes a discount factor.

The choice of a discount (or compound) rate is informed by the theoretical literature and in some Member States there are official rates.

4.5.2 *Determining the Benefits of Primary Remediation*

If primary remediation actions are, or will be, implemented, the benefits, or service gains, of the primary remediation should be calculated. This will entail determining the amount of service improvement, as well as the rate of improvement. Quantification of the benefits of primary remediation could entail consideration of the following factors:

- Comparison to similar primary remediation actions undertaken elsewhere;
- Models of environmental improvement;
- Ecological succession and the time required for ecosystem recovery following disturbance;
- Biological regeneration times; and
- Physical, chemical, or hydrological recovery times.

For example, if the primary remediation involves cleanup and actions to restore the physical environment, then an estimate of the time required for re-establishment of natural functioning of the system could be based on:

- Growth rates or ecological succession patterns;
- Information from the literature on the time required for re-establishment of nutrient cycling, steady-state biomass or habitat structure similar to baseline conditions, or expected biotic community structure or population density; and
- Information on the fate, degradation, dilution, binding and burial, or other elimination or detoxification routes for chemical contaminants should be considered when evaluating the benefits of primary remediation in the case of spills and releases. Information on efficacy of eradication of biological agents should be considered in the case of introductions or releases of non-native species or pathogens.

4.5.3 *Determining Recovery Rates*

The time course of recovery following primary remediation could be either a linear function (e.g., steady recovery after remediation actions are complete, with a monotonic increase in conditions provided each year toward the baseline or expected final state), or a nonlinear function, if data are available to describe such a trajectory.

For some ecosystems, initial recovery from a state of complete destruction to a marginally functional system may be rapid. However, full recovery of function might take many years even after the habitat is visually similar to the baseline conditions. Such a trajectory might be described in linear segments, where the first segment has a steep recovery slope, followed by a more gradual slope for future years. If data are available to describe a more complicated trajectory, such a model can readily be incorporated into an equivalency analysis.

4.5.4 *Consideration of Collateral Damage*

If an emergency response action or the primary remediation causes damage that is additional to the damage caused by the incident, this too should be included in the calculation of loss. Sometimes such collateral damage is unavoidable and necessary to prevent the spread or limit the severity of the incident. For example, responding to an oil spill may require the transport of heavy machinery into an ecologically sensitive area, thus resulting in additional damage that is collateral to the original incident.

4.6 Illustrative Examples of Debit Calculations

4.6.1 *Habitat Equivalency Analysis*

In this Section, we provide an example to illustrate debit calculations using an HEA approach. We developed a simple damage incident with hypothetical assumptions to which we can apply a service-to-service approach.[6] In our simple HEA, we assume that 100 ha of land were damaged leading to a loss of habitat-related functions or services. Below we identify the hypothetical assumptions for our example, which is modelled after the inputs described above.

- *Start year.* We assume losses begin in 2012.
- *End year.* We assume losses accrue until 2021, at which point the habitat functions or services provided by the 100 ha return to the baseline (pre-incident) level.
- *Base Year.* In order to discount, we select a base year (or year in which the values are measured) of 2012, which means the present value multiplier equals one in that year.
- *Spatial extent.* We assume losses occur evenly to the entire 100 ha parcel.
- *Degree of loss.* We assume a 50% service loss based on a decline in our quantification metric: number of species present at the site. In this simple example, we assume that the number of species at the site is a proxy for measuring the level of habitat functions or services. We further assume that the 50% loss persists for the first five years (until 2016) and the loss diminishes each

[6]This example is simple in that it assumes the lost services recover naturally (without the need for primary remediation). Thus, we do not have to account for the benefits of primary remediation or any collateral damage.

year for four additional years, at which point the percent service loss is zero and the system has recovered.

- *Present value multiplier*. To represent a social time preference, we select a 3% discount rate.
- *Metric*. To provide an integrated sense of cumulative habitat functions or services that can be estimated based on expert judgment using metrics like population of important species, vegetation cover and so on.

Table 4.1 displays these calculations. Column (A) shows the spatial extent (i.e., the area impacted); column (B) shows the degree of service loss in percentage; and column (C) shows the present value multiplier, which is based on a 3% discount rate. The annual debit (D) is calculated by multiplying these three columns together. Summing them gives the total debit during the time the service is damaged. The total HEA debit for this hectare is calculated in discounted DSHaYs.

We will use this total debit of 319.5 DSHaYs later in Chap. 6 to illustrate how to scale offsetting remediation for this simple example (see Sect. 6.3.3).

Table 4.1 Illustrated example of debit calculations using a non-monetary metric

Year	Spatial extent (ha)	Degree of loss (% decrease in species on site)	Present value multiplier[a]	Debit[b] (DSHaYs)
	(A)	(B)	(C)	(D) = (A) × (B) × (C)
2012 (base year)	100	50	1	50.00
2013	100	50	0.97	48.50
2014	100	50	0.94	47.00
2015	100	50	0.92	45.76
2016	100	50	0.89	44.42
2017	100	40	0.86	34.50
2018	100	30	0.84	25.12
2019	100	20	0.81	16.26
2020	100	10	0.79	7.89
2021	100	0	0.77	0.00
	Total Debit in 'Discounted Service Hectare Years' or DSHaYs			**319.5**

[a]Present value factor = 1/(1 + discount rate)$^{(\text{year}-\text{base year})}$, where discount rate is 3% and base year is 2012

[b]Debit is calculated by multiplying spatial extent times percent service loss times present value factor

4.6.2 Value Equivalency Analysis

The total debit calculation is identical for both value-to-value and value-to-cost analysis—the only difference between the two analyses lies in the scaling of remediation (discussed in Chap. 6). In this simple value equivalency analysis, we assume that a popular fishing river was contaminated by a chemical release, which led to a complete loss of some fishing trips and a partial loss of other fishing trips over a three year period. Below we identify the specific assumptions for our illustrative example, using the inputs described above:

- *Start year.* We assume losses begin in 2012.
- *End year.* We assume the losses end in 2014, i.e., recovery to the baseline occurs in 2014.
- *Base year.* We select 2012 as the base year, which means the present value multiplier equals one in that year.
- *Number of fishing trips lost.* We estimate that 600 recreational fishing trips will be lost, i.e., not taken, due to the contaminated river over a three year period. That is, 200 anglers who would have fished on this river each year for the next three years will not go fishing at all,[7] thus leading to a welfare loss for these fisherman.
- *Use value of fishing trip lost.* We assume the 'per trip value' associated with these *lost* trips is €25.
- *Number of fishing trips diminished.* We also assume that 100 fishing trips are still taken to the contaminated river each year for the three year period, but that the experience for these fishermen is diminished compared to the fishing experience they could have received prior to the spill. This might manifest itself in terms of fewer fish caught per trip, or smaller fish caught per trip, or a fish consumption advisory, i.e., contamination levels that make the fish dangerous to eat.
- Use value of fishing trip diminished quality. The 'per trip value' associated with these *diminished* trips is assumed to be €15.
- *Present value multiplier.* To represent a social time preference, we select a 3% discount rate.

Table 4.2 displays these calculations. The top half of the table calculates the service loss associated with fishing trips that were not taken (€14,567); the bottom half calculates the service loss due to the diminished experience associated with fishing trips that were taken (€4,370). The sum of these two losses represents the total welfare loss, i.e., lost human use, associated with the spill (€18,938). The

[7]In our simple example we assume the fisherman stay at home. More complex cases may involve traveling to another, less preferable site and the costs (losses) associated with this.

Table 4.2 Illustrative example of debit calculations using a monetary metric

Year	Number of fishing trips lost	Value of fishing trip lost (€)	Present value multiplier[a]	Debit (DLV) (€)[b]
	(A)	(B)	(C)	(D) = (A) × (B) × (C)
2012	200	25	1	5,000
2013	200	25	0.97	4,854
2014	200	25	0.94	4,713
2015	0	25	0.92	0
2016	0	25	0.89	0
Total discounted value of lost trips (€)				**14,567**
Year	Number of fishing trips with diminished quality	Value of fishing trip with diminished quality (€)	Present value multiplier[a]	Debit (DLV) (€)[b]
	(A)	(B)	(C)	(D) = (A) × (B) × (C)
2012	100	15	1	1,500
2013	100	15	0.97	1,456
2014	100	15	0.94	1,414
2015	0	15	0.92	0
2016	0	15	0.89	0
Total discounted value of diminished trips (€)				**4,370**
Total discounted value of lost services (DLV) (€)				**18,938**

[a]Present value factor = $1/(1 + \text{discount rate})^{(\text{year}-\text{base year})}$, where discount rate is 3% and base year is 2012

[b]Debit is calculated by multiplying number of fishing trips lost/with diminished quality times value of fishing trip lost/diminished times present value factor

annual debit is calculated by multiplying columns A, B, and C together. Summing the two losses gives the total debit during the time the human use services were damaged. Note that the losses occurring in the future are worth less in present value terms due to the present value factor in column (C). The total debit for this damage is calculated in discounted lost value (DLV) and measured in a monetary metric of €18,938. We will use this total debit of €18,938 later in Step 4 to demonstrate how to scale offsetting remediation for this simple example (see Sect. 6.3.4).

References

Babut, M. P., Wolfgang, A., Graeme, B. E., Camusso, M., de Deckere, E., & den Besten, P. J. (2005). International overview of sediment quality guidelines and their uses. In R. J. Wenning (Ed.), *Use of sediment quality guidelines and related tools for the assessment of contaminated sediments*. Pensacola: SETAC.

CCME. (1999). *Canadian environmental quality guidelines*. Winnipeg, MB: CCME, Canadian Council of Ministers of the Environment (CCME).

Crane, M. (2003). Proposed development of sediment quality guidelines under the European water framework directive: A critique. *Toxicology Letters, 142*(3), 195–206.

Efroymson, R. A., Will, M. E., & Suter, G. W. (1997a). *Toxicological benchmarks for contaminants of potential concern for effects on soil and litter invertebrates and heterotrophic process: 1997 revision*. Washington: US Department of Energy.

Efroymson, R. A., Will, M. E., Suter, G. W., & Wooten, A. C. (1997b). *Toxicological benchmarks for screening contaminants of potential concern for effects on terrestrial plants: 1997 Revision*. Washington: US Department of Energy.

GeneWatch UK. (2005). *Notes for Defra in relation to GMOs and the implementation of the environmental liability directive*. UK: GeneWatch.

Hin, J. A., Osté, L. A., & Schmidt, C. A. (2010). *Guidance document for sediment assessment*. Methods to determine to what extent the realization of water quality objectives of a water system is impeded by contaminated sediments. Netherlands: Rijkswaterstaat Centre for Water Management.

Lipton, J., Galbraith, H., Burger, J., & Wartenberg, D. (1993). A paradigm for ecological risk assessment. *Environmental Management, 17*(1), 1–5.

Long, E. R., Field, L. J., & MacDonald, D. D. (1998). Predicting toxicity in marine sediments with numerical sediment quality guidelines. *Environmental Toxicology and Chemistry, 17*(4), 714–727.

MacDonald, D. D., Ingersoll, C. G., & Berger, T. A. (2000). Development and evaluation of consensus-based sediment quality guidelines for freshwater ecosystems. *Archives of Environmental Contamination and Toxicology, 39,* 20–31.

National Oceanic and Atmospheric Administration (NOAA). (1999). *Sediment quality guidelines developed for the National Status and Trends Program*, USA.

Ott, W. R. (1978). *Environmental indices, theory and practice*. Ann Arbor Science Publishers.

Strange, E., Galbraith, H., Bickel, S., Mills, D., Beltman, D., & Lipton, J. (2002). Environmental assessment: Determining ecological equivalence in service-to-service scaling of saltmarsh restoration. *Environmental Management, 29*(2), 290–300.

Chapter 5
Step 3: Determining and Quantifying Remediation Benefits

Joshua Lipton and Kate LeJeune

Abstract The purpose of this step of the equivalency process is to determine and quantify the benefits (credits) of created or improved habitats, resources, or resource services that can be used to offset quantified damage through complimentary or compensatory remediation. The process of determining and quantifying the benefits of remediation generally includes developing project-specific criteria for selecting remediation projects, identifying potential remediation projects based on those criteria, and evaluating the nature, degree, and spatial and temporal extent of benefits. Project benefits typically are quantified using the same quantification metrics as those developed for damage determination. When this is not possible, it may be necessary to use habitat or resource adjustment scalars to align remediation credits with damage metrics. In addition, remediation project-specific natural resource or service recovery pathways should be determined to understand the flow of benefits in the future. Sources of potential uncertainty, including likelihood of project success and sustainability in the future, should be explicitly addressed and incorporated into the analysis. In some cases, uncertainty can be evaluated through specifically designed pilot remediation projects.

Keywords Remediation options · Evaluation of remediation options
Remediation benefits · Credits · Uncertainty

J. Lipton (✉)
Abt Associates, 1881 Ninth St., Suite 201, Boulder, CO, USA80304
e-mail: Josh_Lipton@abtassoc.com

K. LeJeune
Le Moulin Du Siffré, 66730 Rabouillet, France
e-mail: klejeune@icloud.com

J. Lipton et al. (eds.), *Equivalency Methods for Environmental Liability*,
https://doi.org/10.1007/978-90-481-9812-2_5

5.1 Introduction

The third overall step in the equivalency analysis framework involves determining and quantifying the benefits that would accrue from the remediation actions. The general approach to the step entails identifying remediation project options and quantifying the resource or service benefits, sometimes referred to as 'credits,' that would accrue from remediation. Non-monetary metric indicators of benefits might include habitat improvements, increases in the population of a specific resource, or improvements in community composition or diversity indices. In the case of the monetary metric, gains might be expressed in terms of increased human uses of a resource (use values) or increases in the value individuals hold for a resource or service, independent of whether they may use the resource (non-use value). Key steps in this portion of equivalency analysis include the following (Box 5.1 and Fig. 5.1):

- *Identify and evaluate potential remediation options*. In this step, potential complementary or compensatory remediation projects that could be used to offset damages are identified. The following are covered in this Chapter:
  - Typology of potential remediation options;
  - Evaluation criteria for remediation options;
  - Remediation project descriptions; and
  - Evaluation of potential project benefits.
- *Calculate habitat, resource, or service gains (credits)*. For each feasible remediation option, the benefits that would result from implementation are quantified. The following are covered in this Chapter:

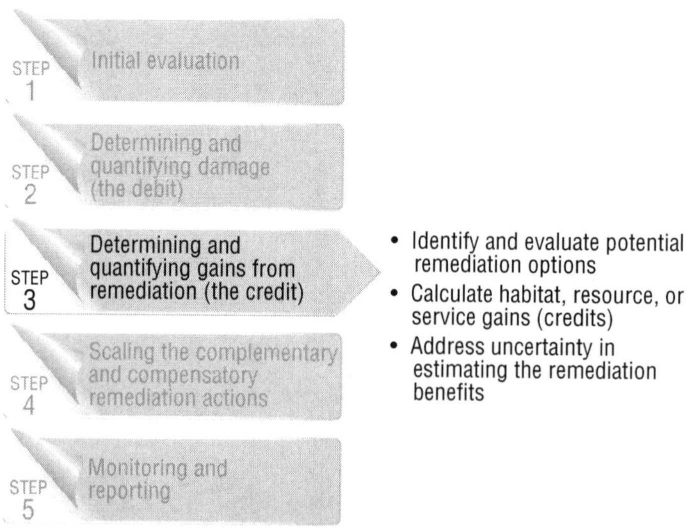

Fig. 5.1 Step 3 of equivalency analysis

  - Determining and quantifying the degree of benefits from remediation; and
  - Determining and quantifying benefit recovery curves.

- *Address uncertainty in estimating the remediation benefits.* The following are covered in this Chapter:
  - Identifying sources of uncertainty, including success likelihoods; and
  - Description, analysis, and incorporation of uncertainty.

Box 5.1: Key issues and actions in Determining and Quantifying the Gains from Remediation

The purpose of **Determining and Quantifying Gains from Remediation** is to identify the extent of ecological or human benefits that would be provided by remediation projects. Key elements of this process include: **identification and evaluation of remediation options**, and **calculation of benefits (credits)**.

During the first part of Step 3, potential projects are screened against a range of selection criteria. Once a set of acceptable projects is identified, the benefits anticipated from each project or project type are identified and quantified in terms of one or more of the metric used on the loss (debit) side of the analysis.

The second part of Step 3 includes the calculation of benefits (credits) of remediation. This requires developing a similar set of information as quantifying damage (debits). Key elements of quantification of the credits are:

- **Determining the degree of natural resource or service improvements over time** in a manner similar to determining the degree of resource or service impairment on the damage side.
- **Determining recovery curves** reflecting the anticipated timing and degree of productivity of the remediation actions evaluated in terms of the chosen metrics.

Throughout an equivalency analysis, analysts often are confronted with **uncertainty**. The calculation of natural resource or service gains of remediation options (degree of improvement, recovery curves, etc.) can be subject to varying degrees of uncertainty.

The final outputs of Step 3 (remediation gains) are used alongside the outputs from Step 2 (total debits) in order to scale remediation (Step 4).

5.2 Identifying and Evaluating Potential Remediation Options

Identification of potential complementary and compensatory remediation projects should be initiated as information about the nature and extent of damages starts to become known. Early identification of potentially relevant remediation projects will assist in the selection of appropriate equivalency approaches and metrics, and may inform decisions about the degree of primary remediation that should be attempted.

Complementary and compensatory remediation planning should include identifying projects that benefit the types of resources and/or services damaged. This step also should involve determining the mandates, preferences, and goals of the Competent Authorities and other stakeholders. Once a set of candidate project types or projects is identified, the benefits anticipated from projects should be identified and quantified in terms of metrics used on the loss (debit) side of the analysis.

It often is necessary to identify several potential projects or sets of projects, because (1) a single project or project type might not be sufficient to offset the entire debit; (2) some projects may be deemed infeasible, too costly, or otherwise inapplicable; (3) Competent Authorities may wish to distribute restoration benefits across different project types (to reduce failure risk using a 'portfolio' approach), locations, or resources/services.

Remediation project ideas can come from a variety of sources such as the resource management expertise of involved persons, existing regional or resource-specific management plans, or input from relevant experts or stakeholders. Remediation projects may be conducted on-site; at nearby sites; or at locations that are geographically removed from the site of the damage if there is a reasonable resource connection to the damage, if there are distributional impact or environmental justice considerations, or if administrative, legal, engineering, or biological factors arise.

In some cases, Competent Authorities may identify very specific remediation projects with known designs and project locations. In other cases, however, it may be more appropriate for Competent Authorities to identify and select project *types* (e.g., wetland restoration, species protection, aquatic habitat improvements), with design and implementation specifics developed following resolution of liability claims.

The process of identifying and evaluating remediation options can include the following steps:

1. Establish evaluation criteria for remediation options;
2. Develop a list or database of potential remediation options;
3. Apply evaluation criteria to identify potential remediation actions;
4. Ensure that appropriate metrics can be used to compare remediation gains (credits) with losses due to damage (debits); and
5. Estimate unit costs for priority remediation actions. Costs should account for the implementation and administration of the action, as well as operation, maintenance and monitoring expenditures required to ensure that the project provides the benefits incorporated in the equivalency analysis.

5.2.1 Typology of Potential Remediation Options

The appropriateness of a remediation project will depend on several variables, including the type of habitats or species populations impacted, the nature of resource damages, and the wider pressures acting on targeted habitats, resources, or services. Although it is not possible to define particular remediation projects here, the following six fundamental principles should be considered when selecting appropriate projects (these are described further below):

- Habitat/species enhancements, rehabilitation, restoration, or re-creation;
- Habitat/species fragmentation and isolation;
- Habitat/species designation and protection;
- Differences between habitat and species compensation;
- Multiple species compensation and remediation; and
- Guidance on *ex ante* compensation for damage to Natura 2000 sites.

Habitat/species enhancement, rehabilitation, restoration, or re-creation
Remediation for damage or loss of habitats or other resources might be achieved by creating new or replacement habitat, rehabilitating or enhancing existing habitats, or otherwise restoring or improving ecological functions. In most instances, the habitat to be remediated should be of a similar type as that which was damaged. Habitat types are defined by Annex I of the Habitats Directive (HD) and European Commission (2007). In some cases, it might be appropriate to restore other habitats that are not of the same type as that which has been damaged. This is the case when restoration of this other habitat helps the ecological functioning of the wider ecosystem of which the damaged habitat is a component or when more than one habitat in an ecosystem has been damaged.

In addition to habitat-focused remediation projects, Competent Authorities may consider projects designed to address specific resources or services. Categories of such projects could include species or population protections, enhancements, or reintroductions; resource or hydrological management actions; land use management; removal of barriers to migration; and provision of human use services such as access to environmental amenities or provisioning resources such as drinking water.

Factors that should be considered in identifying and evaluation remediation projects include the following:

- ***Rates of change***: The rate of natural changes to potentially remediated habitats, resources, or services should be considered. For example, removal of non-native conifer trees from a deciduous woodland may be highly beneficial to its conservation status in the short term. However, if these trees are not naturally regenerating, they might disappear from the woodland through natural processes over the long term. Thus, an assessment of conservation benefit in the short term needs to be balanced against what is likely to happen through natural processes in the long term.
- ***Additionality***: Care is needed to ensure that actions undertaken to restore habitats, resources, or services create ‘additional’ benefits beyond those that would likely occur outside of Environmental Liability Directive (ELD) remediation. With the current promotion of biodiversity action plans and public funding for nature conservation initiatives from wildlife non-governmental organisations and government agencies, some remediation

projects may be undertaken under authorities or Directives outside of ELD actions. To avoid the potential for double-counting credits, a remediation scheme promoted under the ELD should be assessed to ensure that it would not be otherwise undertaken within a reasonable time scale as a consequence of other initiatives. The HD also places an obligation on Member States to restore favourable conservation status to Natura 2000 sites. It might be argued that restoration projects undertaken to compensate or remediate environmental damage thus are funding work that would be required under the obligations of the HD. If this is the case, it may be necessary to make a judgment as to the rate of restoration with or without the benefit of the remediation project. On a related matter, see eftec and IEEP (2010) for a discussion of habitat banking.

Habitat fragmentation and isolation—Article 10 (Habitats Directive) measures
The approach taken when selecting remediation projects will be greatly influenced by the context of the habitat, resource, or service that has been damaged. For example, if the impacted habitat is a small part of a much larger extent of habitat, remediation projects that simply create or restore another patch of habitat of similar size in a wholly different location may not compensate for the damage to the original large habitat patch, due to the principles of island biogeography.

For example, 10 ha of temperate heathland habitat that is part of a 1,000-ha site may have a greater value to biodiversity than 10 ha of new or restored temperate heathland created in isolation (depending on connectivity, edge-effect, and other ecological processes). Remediation for this type of impact may be better achieved by reconnecting three or four heathland fragments with restored or recreated heathland corridors. Alternatively, and depending on the resources in question, there may be cases in which creation of adjacent habitat may be of less value (or greater susceptibility) than creating habitat networks, provided that necessary connectivity exists between locations.

The need to re-stitch elements of the landscape to restore ecological function in this way is referred to in Article 10 of the HD. This section of the HD asks Member States to 'endeavour' to improve the coherence of the Natura 2000 network by managing features of the landscape that are of major importance for wild fauna and flora.

Habitat designation/protection
A potential remediation option for Competent Authorities to consider is to designate a given area as a new nature reserve, or Natura 2000 site, to compensate for damage to or loss of another location. It should be emphasised, however, that this approach does not in itself provide ecological or biodiversity benefits because no new ecological services are generated. However, if an unprotected habitat is being threatened, the protection afforded to it by designation could provide compensation or remediation in the future. To serve as a valid form of or remediation, the threat to the newly designated habitat should be real, relatively imminent, and quantifiable. The new designation must therefore have some tangible and quantifiable benefits to biodiversity conservation and not be merely an administrative exercise.

Differences between habitat and species remediation

Where *ex ante* plans impact protected species populations, approaches to compensatory habitat provision may be different from those taken when the damage is to protected habitats. In some instances, it might be appropriate to compensate for the loss of or damage to an area of habitat used by protected species as if it were a protected habitat. In many instances, the habitat in which the protected species occurs is also protected.

Species populations are often highly mobile and can range over considerable distances. The compensation for damage to species populations requires an understanding of the ecological requirements of a species at different stages of its life cycle, at different times of the year, and even at different times of the day.

Although many protected species populations are associated with protected habitats, there are others that utilise common and unprotected habitats (e.g., cultivated farmland) for part or all of their life cycle. In these instances, there may be greater flexibility in the type of remediation options available. The design of remediation projects to offset damage to species populations can incorporate more direct intervention options, such as providing food for wintering bird populations by spreading grain or cultivating crops upon which birds will feed.

Conservation of migratory species can also consider remediation measures that help the population at a stage of its life cycle that is different from that during which the environmental damage occurred. For example, investigation into damage to the wintering habitat of a population of migratory birds may show that it is the birds' breeding habitat or staging sites on migration routes that are more critical to their survival. In such circumstances, it may be more beneficial to the population as a whole to implement projects that will improve the habitat on the migration routes or in the nesting habitats, if these are thought to be limiting factors. A similar situation may arise with migratory fish, such as the Atlantic salmon (*Salmo salar*). Damage to a reach of a river used for spawning may be compensated for not only by providing primary remediation where the damage occurs but also by improving migration routes (e.g., through the removal of barriers such as weirs) or by improving conditions in estuaries (e.g., improving water quality or reducing the impact of commercial fisheries). Box 5.2 provides an overview of transboundary remediation in cases like these.

Box 5.2: Transboundary remediation

The identification of suitable sites for the provision of remediation requires careful consideration of a number of trade-offs. Where possible or appropriate, remediation sites should be proximate to the damaged site and be of similar habitat type, as this contributes to the maintenance of maximum ecological continuity and provision of environmental benefits. However, in the ecologically fragmented landscape of much of Europe it may not be feasible to find suitable sites for remediation proximate to the damaged site. In other instances, the damaged habitat may be intrinsically rare and hence the location of similar sites for remediation may be some distance away.

In certain circumstances, remediation sites may be best located some distance from the damage site and even across national boundaries. The use of trans-boundary sites for biodiversity remediation raises a number of issues relating to:

- Migratory species;
- Biogeographic regions;
- Habitat fragmentation and ecological networks; and
- Environmental services and use values.

The location of sites for provision of compensation under Article 6(4) of the HD is referred to in two important publications from the European Commission (2000, 2007). These documents emphasise the importance of compensatory sites maintaining the coherence of the Natura 2000 network. In summary, they suggest that compensatory measures proposed for a project should:

- Address, in comparable proportions, the habitats and species negatively affected;
- Concern the same biogeographical region in the same Member State; and
- Provide functions comparable to those which had justified the selection criteria of the original site.

The distance between the original site and the place of the compensatory measures therefore is not necessary an obstacle to remediation, as long as it does not affect the functionality of the site and the reasons for its initial selection. However, the requirement for compensation to be within the same Member State may not be so easily accomplished, especially in sites that cross national boundaries or are in small Member States.

Migratory species

Where environmental damage affects migratory species, such as birds protected under the Wild Birds Directive (WBD) or other species listed in the annexes to the HD, in theory there may be opportunities to provide remediation for this damage at a number of locations along the migration route of the species concerned. However, selection of such locations requires a detailed understanding of the migratory behaviour of the species concerned and the careful identification of any ecological bottlenecks or stressors along these routes. If this level of understanding is available, it may be possible to remediate loss of wintering habitat with improvements to habitat used as staging sites along the migration route or even at breeding sites.

Many overwintering migratory wetland birds form assemblages of species that may have arrived at a wintering site along a variety of migration routes and from a number of distinct breeding sites. As a consequence, remediation for this assemblage of migratory species may not be possible away from the wintering sites. However, many migratory wetland birds utilise networks of wintering sites within a wetland complex and it may be possible to define the extent of this complex of sites and provide remediation at one or more of these so that the overall ecological habitat function is maintained.

Where migratory fish are concerned, removal of obstacles to migration along a river, improvements to the environmental quality (e.g., water and sediment quality), or reduction of predation/fishing pressure at certain points along the migration route may all provide remediation options.

Biogeographic regions

The biogeographic regions of Europe are defined through the HD and form ecologically coherent areas with shared ecological characteristics. Remediation of damage to a European protected habitat and species listed in the annexes to the HD must take place within the same biogeographic region as the damaged site.

Habitat fragmentation and ecological networks

For many species and habitats, ecological function can only be maintained if there is a sufficiently large habitat or there are functional links between habitat patches that form ecological networks or support species meta-populations. Article 10 of the HD requires that Member States take measures to conserve stepping stones and linear features of the landscape that maintain ecological connectivity. Where damage occurs to sites that form part of an ecological network of support for a larger meta-population of species it is important that this is taken into account in the selection of remediation measures. As a consequence, it may be preferable to restore habitat patches across a national border than to select a site within the same Member State where it does not have such ecological function. This may be particularly important in mountain ranges and coastal sites where many large and ecologically coherent sites straddle national boundaries.

Environmental services and use values

The ELD requires that the environmental services provided by a damaged site are also taken into account in the selection of remediation options. For many sites, similar environmental services can only be provided in sites that are within the same Member State and within a relatively short journey time from the damaged site. As the distance between the damage site and the remediation site increase, the human population affected by the environmental damage has to travel further to gain the same environmental benefit. To offset this geographical displacement would require the provision of more remediation, perhaps through the provision of a greater area of habitat.

The influence of geographic location of remediation sites on people's preference was tested in the BABE Forest Fires case study (Chap. 11). This was done through surveying people to determine their preferred remediation option for damage to Spanish black pine forests from accidental forest fires near Barcelona. Intuitively, you might assume that that off-site remediation taking place further away from the damaged site requires more remediation than if the remediation was to take place in a closer location. The distance was not found to be a significant factor within Catalonia but if the remediation was offered outside Catalonia, it needed to be 63% larger to be acceptable to the respondents.

Multiple species compensation and remediation
Environmental damage may impact a number of different species. For example, damage to a river may affect populations of several species of protected fish, as well as bird or mammal species that prey on the fish. Some of these may be well studied and have good baseline data, while others may be much less well studied and may have very different life-cycle requirements. Developing a remediation package that adequately benefits multiple species (or habitats) should be considered.

Guidance on ex ante compensation for damage to Natura 2000 sites
In *ex ante* damage situations (HD, WBD, and Environmental Impact Assessment Directive (EIAD)), plans or projects that cause damage to Natura 2000 sites can only be permitted under certain specified conditions that relate to the absence of alternatives and overriding public interest (see HD Article 6(4)). In such circumstances, Member States are required to take all compensatory measures necessary to ensure the overall coherence of Natura 2000 is protected. Compensation in such situations is therefore related to the protection of the coherence of the network.

In the guidance document from the European Commission on Article 6(4) of the HD (92/43/EEC) (European Commission 2007), the concept of compensatory measures in the frame of Article 6(4) is clarified as follows:

> The compensatory measures constitute measures specific to a project or plan, additional to the normal practices of implementation of the 'Nature' Directives. They aim to offset the negative impact of a project and to provide compensation corresponding precisely to the negative effects on the species or habitat concerned. The compensatory measures constitute the 'last resort'. They are used only when other safeguards provided for by the Directive are ineffectual and the decision has been taken to consider, nevertheless, a project/plan having a negative effect on the Natura 2000 site.

There is no definition of what constitutes the coherence of the Natura 2000 network, but guidance (European Commission 2000, 2007) suggests that compensatory measures should:

- Address, in comparable proportions, the habitats and species negatively affected;
- Concern the same biogeographical region in the same Member State; and
- Provide functions comparable to those that had justified selection criteria of the original site.

The distance between the original site and the place of the compensatory remediation measures is not necessarily an obstacle to remediation, as long as it does not affect the functionality of the site, its role in the geographical distribution, and the reasons for its initial selection (European Commission 2007).

Additional factors that should be taken into account include:

- Likelihood of remediation techniques working;
- Time lag between remediation and provision of benefits; and
- Habitat type/vegetation composition of compensatory habitat.

Compensatory remediation measures focused on habitat enhancement or restoration should therefore provide replacement habitat in 'comparable proportions' to that which are lost or damaged. If there is a geographical displacement between the impacted site and the compensatory site, then additional compensatory habitat may be required. This is particularly important if the damage causes a loss of a large and ecologically coherent site and the compensatory habitat is fragmented and isolated from the damaged site. Inevitably, the compensatory habitat will not have identical ecological characteristics to the damaged site. For example, it may consist of slightly different vegetation communities in different proportions and in different states of ecological condition. Additional compensatory habitat therefore may be required to offset these differences between damaged and compensatory habitat.

Certain types of habitat creation may of necessity be experimental in nature, requiring implementation of new or relatively untested management or engineering techniques. For example, inundation of reclaimed land with seawater might be predicted to create new intertidal habitat. However, the proportions of mudflat, sand flat, and salt marsh within this new habitat and their ecological function in terms of biological productivity, might not be accurately predicted. Additional compensatory habitat might therefore be required to offset this ecological uncertainty.

Regarding interim losses, guidance from the Commission (European Commission 2000, 2007) suggests that the results of the compensatory measures of the HD need to be operational before or at the same time as the damage is caused. Hence, in this scenario no interim losses will be suffered (and have to be compensated for). However, it is not always possible to ensure compensatory habitat has reached its full ecological function before the damage occurs. In such circumstances, additional compensatory habitat should be provided to offset this time lag. Under certain conditions, Member States may take compensatory measures at the same time as the project is being implemented. As compensation takes time to generate benefits, this mismatch in timing may lead to interim losses. One way to compensate for such interim losses is to create a habitat that is larger than the one that was lost. Given the objectives of the HD and WBD, that is, primarily nature protection and conservation, the extra compensatory measures to be taken do not aim, at least not directly, at compensating the interim loss of services to humans suffered due to the fact that the compensatory measures were not in place before the Natura 2000 site was damaged. This is an important difference with the ELD, which clearly stipulates that such losses are to be taken into account (see Article 2(13) and paragraph 1(d) of Annex II of the ELD).

5.2.2 Evaluation Criteria for Remediation Options

The potential remediation options that have been identified should be evaluated with respect to the requirements or preferences of the Competent Authorities and other stakeholders and the relevant statutes and regulations. Annex II of the the ELD explicitly mentions a number of criteria for evaluating reasonable remedial options using best available technologies.

These are:

- Effect on public health and safety;
- Cost of each option;
- Likelihood of success of each option;
- Extent to which each option will prevent future damage and avoid collateral damage as a result of implementing the option;
- Social, economic, and cultural concerns and local factors;
- Time needed for restoration of damage;
- Extent of restoration of the damaged site; and
- Geographical linkage to the damaged site.

As analysts develop and select project evaluation criteria, they may define a wider selection of criteria and describe how each can be interpreted when evaluating proposed projects. Table 5.1 provides examples of criteria and how each criterion is interpreted. Criteria and their interpretations can vary between sites, depending on site-specific issues, opportunities, and constraints. Please note that although the wording used in Table 5.1 refers largely to ELD, the criteria apply equally well to situations covered by the HD, WBD, and EIAD.

The criteria presented in Table 5.1 are divided into two groups: initial screening criteria and detailed evaluation criteria. Criteria such as those in the initial screening group may be used as 'pass/fail' criteria to rapidly eliminate unsuitable options in an objective manner. This can be important if interested parties have contributed ideas that, although potentially meritorious, are not appropriate for offsetting the types of damages incurred.

Table 5.1 does not present a definitive and exhaustive list. Depending on the preferences of the parties involved, different criteria or categories of criteria could be added, removed, and weighted differently to emphasise project characteristics of greatest value to the parties. Weighting can be quantitative, or it can be qualitative if the parties involved conceptually agree on the relative importance of various aspects of proposed remediation projects. More detailed criteria may be developed for use in specific cases. For example,

- *Focus criteria* reflect Competent Authority objectives and preferences regarding the remediation, enhancement, and conservation of the environment. Different parties involved may have very different views about focus criteria, based on site-specific factors of the incident.

Table 5.1 Example evaluation criteria for selecting remediation options

Criterion	Interpretation
Initial screening criteria	
• Address resources damaged by releases, or services lost as a result of damages	• Projects are evaluated with regard to whether they restore, rehabilitate, replace or acquire the equivalent of damaged natural resources and services
• Comply with applicable/relevant laws and regulations	• Projects must be legal, legally implemented, and consistent with applicable regulations
• Protect public health and/or safety	• Projects must not jeopardise public health and/or safety
• Coordinate with planned clean up and primary remediation actions	• Projects must not conflict with planned clean up and primary remediation actions and will not be undone or harmed by these actions
• Be technically feasible	• Projects should have a high likelihood of success
• Minimise collateral damage	• Projects should minimise additional natural resource damage, service loss, or environmental degradation. Collateral damages that may be caused should be limited compared to benefits achieved. Collateral damages should be quantified and factored into benefits calculations
• Be acceptable to the public	• Projects must meet a minimum level of public acceptance; projects should not create a public nuisance
• Reduce exposure of natural resources to contaminants	• Primary remediation projects should reduce exposure to contaminants and reduce the volume, mobility and/or toxicity of contaminants
• Reduce the volume, mobility, and/or toxicity of contaminants	
Detailed evaluation criteria	
• Restore or preserve the type of natural resources damaged	• Projects should improve the quality of the resource that was or will be damaged (e.g., groundwater, terrestrial habitat) through remediation or preservation actions
• Preserve threatened natural communities that are unique, of high quality, or connected to such areas	• Projects that involve land/resource acquisition, protection, or conservation should protect high quality or unique resources or 4 establish viable buffers against future development around such areas
• Target a resource or service that is unable to recover, or that will require a long time to recover naturally	• Projects may target resources/services that will be slow to recover without remediation action
• Address remediation of 'preferred' resources or services	• Resource managers may develop a list of priorities based on the resource types injured and degree of damage

(continued)

Table 5.1 (continued)

Criterion	Interpretation
• Use established, reliable methods/ technologies known to have a high probability of success	• Projects should use appropriate, proven and successful techniques. Experimental methods, research, or unproven technologies may be given a lower priority, or may require implementation of pilot projects prior to full implementation
• Have low costs associated with long-term operation, maintenance and monitoring	• Long term costs of operations, maintenance, or monitoring should be reasonable given the benefits expected
• May be scaled to appropriate level of resource damage or loss	• Projects should be amenable to scaling to provide remediation of appropriate magnitude. Small projects that provide only minimal benefit relative to the damaged resources or services, or overly large projects that cannot be appropriately reduced in scope may not be favoured
• Provide benefits that can be measured to evaluate success	• Projects should produce benefits that are quantifiable through monitoring, modelling, or reasonable extrapolation
• Be consistent with regional planning and be administratively feasible	• Projects should be consistent with regional planning (e.g., supportive of Biodiversity Action Plans); projects must be administratively feasible
• Enhance the public's ability to use, enjoy or benefit from the environment	• This may be considered either as a separate evaluation criterion or a part of collateral benefits, depending on the goals of the parties involved
• Aim to achieve environmental equity and/or environmental justice	• The environmental equity or justice of a project is the degree to which the project benefits the individuals most affected by the damage. Projects that benefit low-income or underserved segments of the human population, or otherwise support environmental justice goals may be favoured
• Provide benefits sooner	• Projects that will achieve full expected results sooner than will be achieved through natural recovery, and sooner than other projects that benefit the same resource, may be favoured
• Provide long-term benefits	• Projects that will persist may be favoured over short-term projects
• Provide benefits not being provided by other remediation projects	• Avoid projects that are already being implemented or have planned funding under other programs, in order to ensure that benefits are supplemental

- *Implementation criteria* are intended to identify relative differences among proposed remediation projects with respect to the costs and ease of implementation for the proposed actions.
- *Benefits criteria* are intended to identify relative differences in the types, timing, scale, and permanence of a remediation project's anticipated benefits. These criteria may also incorporate or reflect the goals of the Competent Authority in terms of preference for certain types of actions or resource services being provided.

Once unacceptable options are eliminated and remaining options are ranked in either a quantitative or qualitative manner, preferred projects (or project types) can be identified. These preferred projects can be given the additional attention needed to prepare more detailed project descriptions, as necessary, and to evaluate potential benefits, including the type, degree, and timing of benefits and the relative fit of the project in terms of its overall equivalency.

5.2.3 Remediation Project Descriptions

The information listed above is used for three purposes: (1) to identify appropriate, highly relevant, and beneficial projects; (2) for equivalency modelling to scale the project such that it offsets the damage; and (3) to prepare a final remediation plan and guide for implementation.

Project descriptions may initially be narrative, single-sentence descriptions. As more information is gathered, project descriptions should become more detailed and should address the following:

- Project objectives;
- Actions required to implement the project;
- Potential project size;
- Anticipated benefits and the time estimated to achieve the benefits;
- Ongoing operation and maintenance activities required to sustain the project or its benefits;
- Approximate cost required for implementation and ongoing operation and maintenance;
- Permitting requirements;
- Potential administrative (or other) obstacles;
- Potential collateral/ancillary benefits or damages associated with the project; and
- A monitoring and evaluation plan including, as appropriate, an adaptive management approach as new monitoring information or scientific data become available.

To assist in describing projects, it may be necessary to develop conceptual designs or models; engineering designs; revegetation plans; and hydrological, geochemical, or biological models, and to investigate similar projects to acquire sufficient information to scope a potential project. Because the project description or potential size could change as more information is compiled, this step is often iterative.

5.2.4 Evaluating Potential Project Benefits

Projects are typically suggested because they address a resource or service that was or will be damaged by an incident. A project might target remediation of the same type of habitat that was damaged, but in a different location. If the habitats are similar, the benefits associated with the remediation project (e.g., provision of food, thermal and hiding cover, nesting and rearing habitat) often are assumed to be similar to the services lost as a result of the damage. However, it may not be feasible to identify remediation projects that address habitats, resources, or services that are identical to those lost. For example, the remediation site might have a different landscape context, so the ecological benefits may not be identical. The habitat at the remediation site might be more or less accessible to target species or it might be bordered by more or less protective and desirable habitat types. Or, the stressor that has degraded the remediation site may be different from the stressor that caused the damage being offset, so even though ecological benefits are possible and beneficial at the remediation site, they may not be identical to the those required for the offset.

These types of considerations are taken into account in the evaluation of potential project benefits. Sometimes a metric that allows analysts to quantify differences in the quantity of a natural resource or service, or the quality of natural resources or services, provided by the remediation site can be selected. A metric that takes into account multiple attributes of a habitat can be useful. However, for most projects that involve creating or enhancing ecosystem benefits, full quantitative estimates of all anticipated benefits and the time to achieve benefits may not be feasible. Data from the scientific literature, ecological models, and best professional judgment may be used to estimate benefits and ecosystem recovery rates.

Another approach to evaluating relative project benefits across habitats is to use 'scalars' to assign preference weights. Scalars can be used to account for 'preferred habitats' (e.g., some highly productive wetland habitats may be 'preferred' to less productive grasslands); for species or habitat scarcity; for distance from the damaged site (e.g., Competent Authorities may prefer to perform remediation near the incident rather than further away; such distance-related preferences could be reflected in scalars); or for certain social factors.

Below, we discuss three things to consider when evaluating the potential benefits of a remediation project option.

Geographic proximity

Remediation projects that benefit natural resources may be more relevant if they are geographically proximate to the damaged site. Similar habitats and resources may be more likely to be found in a nearby location that shares similar climate, season length, geologic parent material, potential natural vegetation, species assemblages, and natural and anthropogenic stressors. However, it is not always possible to find appropriate restoration projects in close geographic proximity, particularly if the habitat or resources damaged were rare or if the service damaged was dependent on a unique landscape context. In addition, if the damage was to a resource such as migratory birds, and an appropriate remediation project is replacement of migratory birds, then remediation might best be located in a distant breeding ground. As noted above, Competent Authorities may, in certain cases, employ proximity scalars to account for the unavailability of remediation opportunities near the incident location (see Box 5.2 for transboundary remediation).

In compensating for human use and non-use losses, it may not be appropriate to conduct remediation far from the damage site. This risks benefiting a population that was not harmed by the loss, at the expense of individuals who experienced the loss, or not properly accounting for environmental justice considerations.

Other ecological, cultural, economic, and social issues

Other issues that might arise in identifying the benefits of a remediation project include compensating for cultural, social, or economic losses that are difficult to describe or quantify. If Habitat Equivalency Analysis (HEA) or Resource Equivalency Analysis (REA) are used to scale remediation because the main loss is an ecological attribute, but a secondary loss is cultural, then an additional method might be needed to scale adequate compensation, including economic valuation methods that measure the values people hold for services provided by natural resources.

Use of value equivalency analysis

Annex II of the ELD gives preference to resource or service equivalency methods. However, under certain situations, it provides for the Competent Authority to use alternative economic valuation methods to scale complementary and compensatory remediation actions. There are two general categories of Value Equivalency Analysis (VEA): the value-to-value method and the value-to-cost method. The value-to-value method equates the monetary value of the damage caused by contamination to the monetary value of the remediation actions. The value-to-cost method determines the monetary value of the damaged resources and then applies that amount of money to implementation of remediation actions. Below, we explain the basis for the VEA approach (i.e., how ecological resources or services can provide both direct use benefits and indirect benefits to those who value the resource even if they may not use it). For information on economic valuation methods see Chap. 8. We also discuss when VEA may be an appropriate method to use and conclude with an important point about the value-to-cost approach.

Translation of ecological services to human services

The premise of the value equivalency method, as with the HEA and REA methods, is that natural resources provide benefits to the public through the provision of services. Services (or *natural resource services* or *ecosystem services*) refer to the functions performed by a natural resource for the benefit of another natural resource and/or people (see Chap. 2). The economic literature on natural resource assets and ecological services uses an analogy to monetary assets and the flow of payments from those assets. Natural ecosystems are assets that, if managed appropriately, can provide the public with a flow of beneficial services over time. Typically, it is impossible to describe all of the services that an ecosystem provides. Fortunately, to implement HEA, REA, or VEA, it is unnecessary to define all the possible services. Only a few significant services, those that correspond to key functions and the effects of the release, need to be defined.

Human services (and values) can flow either directly or indirectly from the ecosystem (also see Box 2.3 for a description of ecosystem services). Direct services (values) include provisioning services (e.g. fishing for food) and cultural services (e.g. enjoying the view of a clean shoreline or landscape). Indirect services (values) include regulating and supporting services (e.g. the ability of the clean shoreline and near-shore ecosystem to provide for a healthy community of fish that in turn supports a productive fishery). There are also non-use (or passive use) values that flow from ecosystems that are associated with individuals' preferences for others to enjoy the services (altruistic value), for passing on a clean environment to future generations (bequest value) and for the sake of the environment itself (existence value).

When is value equivalence analysis appropriate?

VEA is likely to be most appropriate when the nature, scale, or location of remediation projects differs from the specific resources and services damaged.

Section 1.2.3 of Annex II in the ELD states:

> If it is not possible to use the first choice resource-to-resource or service-to-service equivalence approaches, then alternative valuation techniques shall be used. If valuation of the lost resources and/or services is practicable, but valuation of the replacement natural resources and/or services cannot be performed within a reasonable time-frame or at a reasonable cost, then the competent authority may choose remedial measures whose cost is equivalent to the estimated monetary value of the lost natural resources and/or services.

The ELD provides no more specific requirements or recommendations on when use of the resource-to-resource or service-to-service methods may be inappropriate. Value based equivalency analysis may be more appropriate in situations where the answer to one or more of the following questions is yes:

- Do the damaged resources differ in type from the remediated (complementary or compensatory) resources?
- Is 'in-kind' remediation infeasible, such that different resources or services must be remediated to compensate for environmental harms?
- Do the damaged resources differ in quality from the remediated (complementary or compensatory) resources?
- Is the scope of the damage so large that the necessary assumptions of service-to-service and resource-to-resource equivalency are not supportable?
- Are important human use services lost as a result of the damage?
- Is the location of remediation actions sufficiently far from the damaged site that value equivalency should be considered?

Value-to-cost approach

In the value-to-cost approach, the value of the damage is estimated using a valuation method. However, it may be too difficult estimate the value of the remedial actions. Instead, the Competent Authority takes the amount of money estimated as the value of the damage and selects a remediation option(s) that costs as much as the value of the damage.

In the value-to-cost approach, it is not directly known if the value to the public generated by the remedial actions is equal to the value of the loss. The value may be less or greater than the damage, which is why the value-to-cost approach is reserved as the last option for use by the ELD. Furthermore, the value-to-cost approach crucially depends on the definition of the affected human population. Depending on the type and scale of damage, this could be the local user population only, i.e. those who used the damage of resource in one way or another, prior to the incident and could have continued to use it had the incident not taken place. The affected population could be regional, national, throughout the European Union, or even global depending on the increasing level of damage and uniqueness of the resource. While user population relatively easy to identify, the potentially much larger non-user population may not be identifiable without empirical economic research. As the damage is estimated by multiplying the debit value per person with the number of people affected, the size of the population would result in large differences in the damage value and hence remediation budget. For information on scaling remediation under a value-to-cost framework, please see Sect. 6.3.2.

5.3 Calculating Benefits (Credits) of Remediation Options

Quantifying the benefits of potential remediation projects (credits) requires developing a similar set of information as for quantifying damage (debits). The anticipated timing and degree of productivity of the remediation actions must be evaluated in terms of the chosen metric. The anticipated change in habitats, resources, or services over time then should be quantified.

Figure 5.2 depicts how resource improvements might accrue following implementation of compensatory or complementary remediation actions that improve resource or service conditions. The dashed line represents the increase in resources or services resulting from a remediation action. Once the action ends, resource and service quality might continue to improve to some new baseline, which would persist for some time into the future. The grey area (the area between the baseline and the improvement path) represents the remediation benefits used to offset the losses associated with the damaged site.

5.3.1 Determine the Degree of Benefits from Remediation

The degree of natural resource or service improvements, or credits, is estimated in a manner similar to determining the degree of natural resource losses when

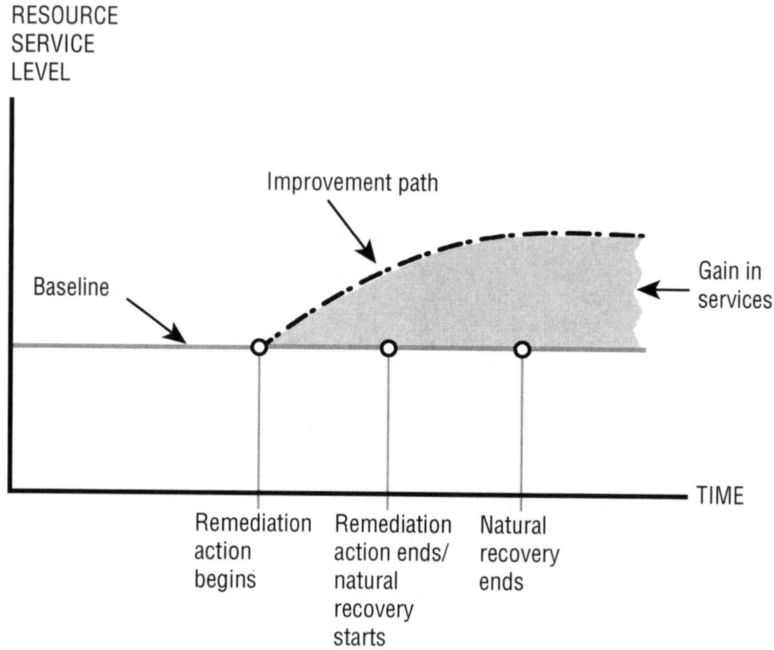

Fig. 5.2 Quantifying anticipated improvements from compensatory remediation

calculating debits. The natural resources or services likely to be provided by the remediation project should be identified. The degree of natural resource or service improvement is calculated by quantifying the change in the selected metric(s) over time following remediation.

5.3.2 Determine Recovery Curves

The anticipated timing and degree of benefit of the remediation actions must be evaluated in terms of the chosen metric(s). An estimation of the amount and future trajectory of benefits can rely on information from similar projects or conditions at similar sites, be based on published literature, rely on the use of computer models, or entail professional judgment.

The time course of benefits accrual following remediation may be described as a linear function (steady increase in natural resources or services provided after remediation actions are complete, with a monotonic increase provided each year to baseline or expected final state), or as a nonlinear function if data or theory support such a trajectory. For example, for some ecosystems, initial responses might be rapid, but attainment of full function might take many years. For simplicity in equivalency modelling, such a trajectory could be described in linear segments, where the first segment has a steep recovery slope, followed by a more gradual slope for future years. Alternatively, if more data are available to describe a more complicated trajectory, such a numerical model could readily be incorporated into an equivalency analysis.

Computer models, such as ecosystem or hydrological models, may be applied to estimate the time required for the remediated resources or habitats to provide full benefits. Natural resource or ecological service provision might be restored at a different rate from human use or non-use service provision. Consequently, both provisions should be considered to determine the amount of remediation necessary to fully compensate for losses, unless a decision has been made to address one or the other in the equivalency analysis.

Determining the time course of recovery might involve discussions with experts such as land or resource managers, interpretation of data from the published scientific literature, and/or comparison to similar projects. The estimate of recovery should take into account natural stressors and disturbances that might reasonably be expected to affect the recovery rate, as well as corrective and maintenance actions that would be taken to bolster project success in future years.

5.4 Dealing with Uncertainty and Variable Outcomes of an Equivalency Analysis

Analysts often are confronted with uncertainty throughout an equivalency analysis. We address uncertainty in this Chapter because the calculation of the benefits of remediation options (e.g. degree of improvement, recovery curves) is often subject to varying degrees of uncertainty.

5.4.1 *Sources of Uncertainty*

Uncertainty associated with equivalency analyses may stem from environmental variability and stochasticity; from measurement uncertainty and variability; from our limited knowledge of ecosystems or ecosystem processes; from a lack of data (or from imprecise data) even when the systems are known and understood; engineering uncertainties; uncertainties regarding the likelihood of success of certain remediation projects; uncertainties regarding future states of nature; or from social, economic, or political decisions, which often involve uncertainty. Uncertainties may be experienced at any stage of the analysis, including when:

- Estimating the losses (debits) from damage and gains from remediation projects (scaling);
- Estimating the benefits (credits) from remediation projects;
- Implementing the remediation projects;

- Addressing administrative, policy and legal matters; and
- Estimating remediation costs.

Natural variation can make it difficult to define and predict recovery trajectories for many habitat services (Strange et al. 2002). The inherent complexity of an ecosystem, whether healthy or damaged, makes the outcome of remediation efforts even more difficult to predict. Although this differs from one equivalency analysis to another, many of the factors that might influence recovery and remediation success may be unknown or poorly understood.

5.4.2 Describing, Analysing and Incorporating Uncertainty

When performing an equivalency analysis, the uncertainty, variability, and probabilities of the possible outcomes associated with an incident falling under the ELD or any other relevant Directive should be considered. Considerations can include:

- Identifying key sources of uncertainty;
- Reducing uncertainties when practical;
- Quantitatively incorporating uncertainties through use of sensitivity analyses, Monte-Carlo simulations, or other numerical tools; and
- Analysing, incorporating, and communicating uncertainties in the presentation of results.

When evaluating uncertainties, reasonable worst-case scenarios may be used to ensure the protection of affected environments and the public.

Early in the process, a value-of-information approach could be used to determine whether to perform additional investigations in order to reduce the sources of uncertainty. A formal or informal value-of-information framework allows consideration of whether the cost of additional studies is warranted given the likely improvement in the accuracy or precision of the final estimate. Such a framework might involve the following:

- List each factor believed to introduce significant uncertainty into damage and remediation benefit estimates;
- Rank these factors according to the magnitude of their effect;
- Determine the extent to which uncertainty in the estimates might be reduced through additional studies;
- Estimate the cost of undertaking additional studies; and
- Identify those studies that are likely to most cost-effectively to reduce uncertainty in the debit and credit estimates to an acceptable level.

Analysis of remaining uncertainties can range from qualitative consideration of the sources, magnitude, and direction of uncertainty, to simple sensitivity analyses, which identify the range of possible risks, to sophisticated probabilistic approaches using Monte Carlo simulation techniques (Metropolis and Ulam 1949; Kahneman and Tversky 1982; Fishman 1995).

Selection of the metric is just as important as the type of sensitivity or Monte Carlo Analyses performed. For example, the choice of a metric, or even the type of data used to quantify the metric, can greatly influence debit estimates because of the inherent variability in the data (recall that selection of the metric influences not only the amount of loss but the spatial and temporal extent of losses and recovery rates). Similarly, metric selection is important in determining the recovery path on the credit side of the equation. Ideally, Competent Authorities evaluate uncertainty by considering how alternative metrics and supporting datasets might influence the nature and extent of remediation in order to ensure that the public's interests are protected (Strange et al. 2002). One common way to do this is to conduct sensitivity analysis on various key assumptions. The idea is to identify the range (difference between the minimum and maximum estimation) of possible remediation estimates and to use this range as a measure of uncertainty. Wider ranges indicate greater uncertainty in the final estimate.

References

European Commission. (2000). *Managing Natura 2000 Sites - The provisions of Article 6 of the Habitats Directive*, Luxembourg.

European Commission. (2007). *Guidance document on Article 6(4) of the 'Habitats Directive' 92/43/EEC*, Brussels.

eftec & IEEP. (2010). The use of market based instruments for biodiversity protection: The case of habitat banking. Report to the European Commission. In K. ten Kate, J. Treweek, & J. Extrom (Eds.), *Collaboration with stratus consulting* (now Abt Associates).

Fishman, G. S. (1995). *Monte carlo: Concepts, algorithms, and applications*. New York: Springer Verlag.

Kahneman, D., & Tversky, A. (1982). *Judgement under uncertainty: Heuristics and biases*. Cambridge: Cambridge University Press.

Metropolis, N., & Ulam, S. (1949). The monte carlo method. *Journal of the American Statistical Association., 44*(247), 335–341.

Strange, E., Galbraith, H., Bickel, S., Mills, D., Beltman, D., & Lipton, J. (2002). Environmental assessment. Determining ecological equivalence in service-to-service scaling of salt marsh restoration. *Environmental Management, 29*(2), 290–300.

Chapter 6
Step 4: Scaling Complementary and Compensatory Remediation

David Chapman and Joshua Lipton

Abstract The purpose of the scaling step of the equivalency analysis is to determine the amount of remediation required to offset damages to natural resources or services. It involves calculating the benefits (credits) for relevant remediation options and determining how much of the selected remediation is required to generate sufficient credit to offset the damage (debit). The determination of how much of the selected remediation is required is called scaling. Estimating the costs of undertaking the necessary amount of remediation options is also discussed.

Keywords Scaling remediation · Remediation credit · Remediation costs

6.1 Introduction

The fourth step in performing an equivalency analysis involves scaling the benefits of remediation such that they offset the quantified environmental damage (debits). This step helps answer the question: 'How much remediation is necessary to compensate for the damage caused as a consequence of the incident?' Key steps in this portion of equivalency analysis include the following (see Box 6.1 for the key issues and Fig. 6.1 for key substeps):

- *Calculate per unit credits*. In this step, service gains of a remediation project are expressed in terms of each unit of service, resource, habitat, or value that is to be remediated, as quantified using the same metric(s) used to calculated debits from the damage.

D. Chapman (✉)
Rocky Mountain Research Station, United States Forest Service,
240 West Prospect Road, Fort Collins, CO 80526, USA
e-mail: davidchapman@fs.fed.us

J. Lipton
Abt Associates, 1881 Ninth St., Suite 201, Boulder 80304, CO, USA
e-mail: Josh_Lipton@abtassoc.com

J. Lipton et al. (eds.), *Equivalency Methods for Environmental Liability*,
https://doi.org/10.1007/978-90-481-9812-2_6

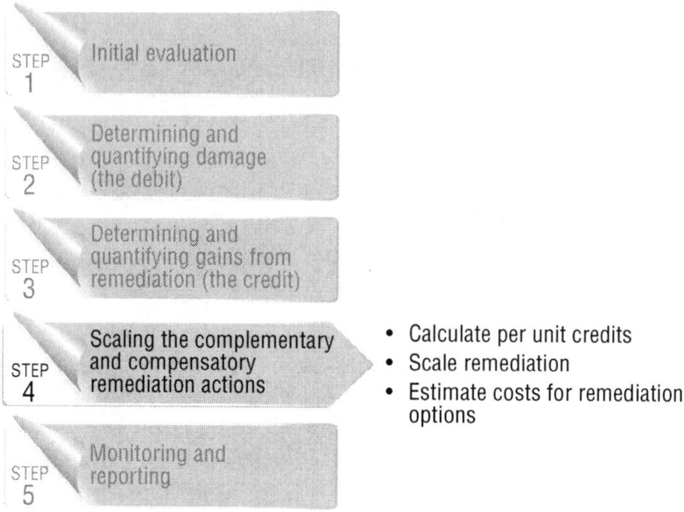

Fig. 6.1 Step 4 of equivalency analysis

- *Scale remediation*. Scaling remediation generally entails dividing the total debits by the per unit credits to determine how much remediation to provide.
- *Estimate costs for remediation options*. When using equivalency analysis, the cost of environmental liabilities includes the cost to implement and maintain the required remediation projects, as well as the cost of the efforts to undertake the equivalency analysis and the costs associated with planning, overseeing, and monitoring the remediation projects.

When scaling is undertaken, the credits from a remediation project should be quantified using the same metric(s) as used for the damage (debit) calculations (Chap. 5) so that the amount of remediation needed offsets the amount or extent of environmental damage.[1]

The general approach to scaling remediation is the same for Habitat Equivalency Analysis (HEA), Resource Equivalency Analysis (REA), and Value Equivalency Analysis (VEA), but the specifics differ slightly. For example, HEA and REA rely on non-monetary metrics (e.g., habitat services, number of resource units), while VEA relies on a monetary metric. However, in all cases, the scaling step is used to determine how much remediation is necessary, using the identified metrics, to offset the damages.

[1]For purposes of simplicity of exposition, throughout this chapter we present the discussion as if one type of remediation project is being implemented that is scalable to offset debits. In practice, multiple different remediation actions may be under consideration for use, and not all remediation projects may be completely scalable. The fundamentals of the approach described in this chapter still apply, however.

Box 6.1: Key issues and actions in Step 4: Scaling the complementary and compensatory remediation actions

The process of determining the amount of complementary and compensatory remediation that compensates for damage and interim loss is called scaling. Because scaling requires balancing the debit and credit sides of the equivalency equation, it generally requires two inputs: (1) the total discounted losses, or *total debits* (Chap. 4) and (2) the per unit present value gains, or *per unit credits*, of the remediation project. Remediation scaling then is achieved by dividing the total debits by the per unit credits, which gives you the appropriate amount (e.g., scaled amount) of remediation needed to offset the damage. This approach works for both monetary and non-monetary metrics. However, in the value-to-cost approach, scaling is simplified: the cost of remediation is set equal to the monetary amount of damage.

Step 4 also provides information on estimating the costs of remediation. The key elements of scaling complementary and compensatory remediation that should be considered in an assessment include the following (summarised briefly below):

- Calculating gains (credits) on a per unit basis for potential remediation options that pass the initial screening;
- Scaling remediation by dividing total debits by per unit credits to obtain the appropriate amount of remediation to offset damage; and
- Estimating costs for remediation options to provide an estimate of the costs of alternative remediation options, which can be useful in comparing across alternatives.

As with steps 2 and 3, uncertainty should be considered in scaling as well.

Estimating gains (credits) on a per unit basis conceptually is very similar to the process of estimating debits. For both monetary and non-monetary metrics, the approach entails identifying key assumptions about how the remediation project is expected to provide gains and then to sum these gains over the life of the project. The formula for estimating the per unit credit differs slightly depending on whether you are applying a non-monetary metric or a monetary metric.

Given an estimate of the per unit credits, the next step is scaling remediation. The formula is quite simple: total debits divided by per unit credits, adjusted on a present value basis. The result is the quantity of remediation, on a present value basis, that compensates for the total discounted present value of the damage. The only exception to this approach is the value-to-cost approach, which does not require an estimate of the (per unit) credits from remediation. Instead, the remediation is scaled based only on the extent of damage (debits). Illustrated examples of scaling remediation are provided for both a non-monetary metric (Sect. 6.3.3) and a monetary metric (Sect. 6.3.4).

Remediation costs include the cost of the remediation assessment, implementation, administration, operation, maintenance, and monitoring. These costs typically are both project- and site-specific.

6.2 Calculating Remediation Gains (Credits)

Although the total remediation benefits of a specific project can be calculated, many types of remediation projects have adjustable sizes, both in terms of geographic/ temporal scope and intensity. For example, the number of hectares of forest revegetation, the intensity of wetland restoration projects or stream habitat improvements, the spatial or temporal extent of bird conservation measures, the amount of water quality improvements, or the number of migratory barriers removed all represent adjustable quanta of remediation alternatives. When a highly specified project hasn't been identified, the quantification of credits per unit of remediation therefore can be an efficient approach to scaling remediation and calculating the amount of remediation that might be undertaken to compensate for losses. *Per unit* credit calculation thus refers to the quantification of the service gains of a remediation project that are expressed in terms of each unit of service, resource, habitat, or value that is to be remediated, as quantified using the same metric(s) used to calculated debits from the damage. If the size or duration of a remediation project is adjustable, the amount of the remediation project can be scaled to fit the extent of damage.

6.2.1 Per Unit Credits: Conceptual Approach with a Non-monetary Metric

The conceptual approach to estimating per unit credits from a remediation option using a non-monetary metric is summarised by the following formula:

$$\sum_{t=0}^{t=n} \frac{(1 \times b_t)}{(1 \times r)^t}$$

where $\sum$ is the summation sign, $t = n$ is the end year, $t = 0$ is the start year, b_t is the degree of gain every year, r is the discount (or compound) rate, and t is any given year in the credit period (between 0 and n).

The inputs for this formula, which are described below, are very similar to those used for the debit formula in Chap. 4. For example, some of the inputs are necessarily the same (e.g., non-monetary metric to measure change, discount rate, and base year); other inputs are very similar in concept (e.g., degree of *gain* on the credit calculation is analogous to the degree of *loss* in the debit calculation). The inputs used in this formula are:

- *Start year* ($t = 0$). The year the remediation project begins providing environmental benefits.
- *End year* (n). The year the remediation project stops providing environmental benefits. In some cases, projects may provide benefits indefinitely. However, it

is still possible to estimate the *finite* benefits provided in such cases through the use of a positive discount rate.

- *One unit (1).* Represents the unit of remediation that can be adjusted, to offset the damage. This may represent a hectare of habitat, a resource such as a fish or bird, etc. In this formula, it is always set to 1 because we are estimating 'per unit' credits.
- *Present value multiplier (r).* As described previously (see Box 4.4), this multiplier adjusts the value of benefits into today's terms. In this case, future benefits are discounted back to present value terms. The discount rate should be the same on both the debit and credit calculations.
- *Degree of gain (b_t).* The degree of gain describes the same concept as the degree of loss in the debit calculation but refers to the *improvement* provided by the remediation project instead of the damage caused by the incident. It is often measured in percentage terms (e.g., percent increase in resources or services) or in number of resource units (e.g., numbers of fish, gallons of water). This rate of change into the future is analogous to the recovery rate on the debit side.

Additional assumptions that are not explicitly in the formula above but are nonetheless important inputs into this calculation include the following:

- *Metric.* The (non-monetary) metric used to measure the gain must be the same as the metric used in estimating the total debits (see Chap. 4 which reviews metric selection), or normalized using an appropriate adjustment scalar.
- *Base year.* The year used for the present value calculations. The year must always be the same as the base year used in the debit calculations.

The illustrative example in Sect. 6.3.3 demonstrates how these calculations might look for a sample HEA. The calculations would be very similar for an REA and therefore are not shown below.

6.2.2 Per Unit Credits: Conceptual Approach with a Monetary Metric

The per unit credits from remediation using a monetary metric are only relevant under the value-to-value framework, where the remediation gains in monetary terms must be quantified in order to scale remediation against losses denominated in a monetary metric. In the value-to-cost framework, the gains from remediation are not scaled through an equivalency analysis to the damage. Below we describe this relevant approach under both frameworks.

The following discussion uses *use* value (as opposed to *non-use* value) as the primary component of losses being valued. Alternatively, the non-use value, or total value, associated with a damaged resource may be compensated for through remediation projects. In these cases, the methods are the same, but the 'degree of gain in human use' would be replaced by the 'degree of gain in non-use or total value' below. (See Chap. 11, the BABE Forest Fires case study, as an example of the substitution of the per unit approach with a direct scaling approach, in which the overall debit value is calculated first and the overall required equivalent credit in environmental terms is estimated directly instead of inferring a per unit value).

Under the value-to-value framework where use value is the primary component, when a single type of remedial action that can be altered in size is being considered, per unit credits are calculated for the scaling process. The conceptual approach to estimating per unit credits from a remediation option using a monetary metric is summarised by the following formula:

$$\sum_{t=0}^{t=n} \frac{(1 \times q_t \times p_t)}{(1 \times r)^t}$$

where $\sum$ is the summation sign, $t = n$ is the end year, $t = 0$ is the start year, q_t is the degree of gain in human use every year, p_t is the degree of gain in economic value per unit of use every year, r is the discount (or compound) rate, and t is any given year in the credit period (between 0 and n).

The inputs are very similar to those used in the total debit calculation in Chap. 4. For example, some of the inputs are necessarily the *same* (e.g., monetary metric to measure change, discount rate and base year), while other inputs are *very similar* in concept (e.g., units of *gain* in human use on the credit calculation is analogous to the units of *loss* human use in the debit calculation). One key difference is the inclusion of the 'degree of gain in economic value'. This calculation is necessary on the credit side when using a monetary metric because it translates changes in resource improvements into the value that people place on that change (in a money measure) associated with that change. Inputs used in this formula include the following:

- *Start year* ($t = 0$). As above.
- *End year* (n). As above.
- *One unit* (*1*). Represents the unit of remediation that can be scaled, that is, adjusted, to offset the damage. In this *use* value example, it can represent a unit of human use (e.g., fishing trip, boating trip, recreational day at a beach). In this formula, it is always set to 1 because we are estimating 'per unit' credits.
- *Degree of gain in human use* (q_t). The improvement associated with human use of a natural resource following a remediation project. For example, if the primary human use is fishing, this may refer to an increase in fishing trips due to an increase in the number of fish caught (or size of fish) at a particular lake following a remediation project (e.g., habitat improvement). Estimating q_t requires knowing the change in the resource or service due to an incident, and what that change means in terms of human use. Where the metric for the former would be ecologically based as discussed in Chap. 4, the metric for the latter will be have to be discernible and hence valued by humans. For example, BOD (Biological Oxygen Demand) may be the correct ecological metric for the initial change in q_t to estimate the effect on fish populations, for the change in human use, the effect of the change in the fish

populations on the relevant recreational activities (if the fish is used for angling) or human health (if it is used a food source) needs to be identified.

- *Degree of gain in economic value* (p_t). The increase in value associated with human use of a natural resource following a remediation project. It translates the degree of gain in human use into an economic gain (measured by our monetary metric) which can be compared to the economic loss (measured by our monetary metric) from the damage. If the primary human use is fishing, this may refer to an increase in the value a fisherman associates with a fishing trip taken following a remediation project at a given location. This link between an increase in a resource measure (fish) to human value could be based on a review of the economic literature describing how fishermen value changes in fishing attributes or through a primary survey.
- *Present value multiplier* (r). As above.

6.3 Scaling Remediation

When scaling remediation, the objective is to determine how much remediation to provide using either a non-monetary metric, and thus HEA or REA, or a monetary metric, and therefore VEA. Below, we provide a description of remediation scaling for each type of equivalency analysis.

6.3.1 Scaling Remediation with Monetary and Non-monetary Metrics

Scaling remediation generally entails dividing the *total* debits by the *per unit* credits.[2] The output is the amount (magnitude) of remediation to provide today (and last for some time into the future[3]) that will offset the damage caused. Thus, a simple formula for scaling remediation is:

$$\text{Total quantity of the remediation project to provide now} \\ = \text{Total present value debits / present value per unit credits}$$

In the case of a non-monetary metric, the number of units of remediation to provide would be the units of habitat, resources, or services that compensate for the damage, measured using the selected non-monetary metric(s).

---

[2]This same process can be applied when credits are not calculated on a unit basis by summing remediation project credits until the full debit is satisfied (see BABE Forest Fire case study in Chap. 11).

[3]How long the remediation benefits will last into the future is an important assumption to be made during an equivalency analysis.

In the case of a monetary metric using a value-to-value approach, the number of units of remediation to provide would be the value associated with the increase in human use (e.g., number of user days)[4] that comes from the remediation project (remember that when using this approach the remediation project must offset the value of the damage). Thus, the Competent Authority and/or responsible operator would need to undertake sufficient remediation to ensure the gain in value is equal to the loss in value. The cost of providing this amount of remediation, including the costs of operations, management, and monitoring, would represent a part of the environmental liability (the other part includes the cost of conducting the equivalency analysis, see Sect. 6.4). Note that this cost in monetary terms may be more or less than the value of the damage, depending on how the users of the resource value the remediation improvement.

6.3.2 Scaling Remediation Under a Value-to-Cost Framework

Scaling remediation under a value-to-cost framework is different. Instead of dividing the total debits by the per unit credits, the amount of remediation to provide is based only on the size of the damage (thus, no need to estimate per unit credits). The scaled amount of remedial actions to ensure equivalence between debits and credits is based on the total damage caused, rather than the value derived from the proposed remediation project (as is required under the value-to-value framework described above). That is, the remediation project is scaled so that its cost equals the total value of the damage. In practical terms, this means that the Competent Authority recovers the full value of the damage and uses these funds to implement a remediation project. Thus, the amount of remediation is scaled based on what it would cost to implement a remediation project that meets the criteria discussed in Sect. 5.2.2.

Note that both the value-to-value and value-to-cost frameworks are equally valid approaches for the purpose of equivalency analysis. The decision to use one or the other will depend upon the desires of the Competent Authority and the responsible operator. However, for damage cases under the Environmental Liability Directive, a specific hierarchy has been established that favours the use of value-to-value over value-to-cost.

---

[4]Depending on the resource, the number of user days may represent fishing trips to a river, number of boating days in a lake, or number of beach visits to a recreational beach. Depending on the type of damage, other units such as health impacts, crop value etc. can also be used.

6.3.3 *Example: Scaling Remediation Using a Non-monetary Metric*

To provide a numerical illustration of how to scale remediation, we use the example presented in Sect. 4.6.1. In that example, we assumed that 100 ha of land were damaged, leading to a loss of functional habitat services. We estimated the total debits to be 319.5 Discounted Service Hectare Years (DSHaYs) (see Sect. 4.6.1 and Table 4.1). For purposes of this illustration, we assume that a remediation project at a nearby location could provide improvements in functional habitat services that are similar to those that had been provided by the damaged land and the habitat improvements (credits) are quantified as described in Chap. 5. Below we identify the hypothetical assumptions for our illustrative scaling example (Table 6.1 summarises the calculations):

- *Start Year*. We assume remediation benefits are first realised in 2014.
- *End year*. We assume benefits from the remediation will stop being provided in 2068.
- *Unit* (*Table* 6.1, *column A*). Hectares of habitat functional services (i.e., unit = hectare).
- *Degree of gain* (*Table* 6.1, *column B*). We assume an ultimate 50% increase in provision of habitat relative to baseline. This gain is assumed to occur gradually in the first five years from 2014 to 2018 and then continue at a constant 50% increase for the next 50 years (at which point the habitat improvements return to the original baseline).
- *Present value multiplier* (*Table* 6.1, *column C*). We assume a 3% discount rate.
- *Metric*. The non-monetary metric is the same as in the debit calculations: hectares of habitat quality function.
- *Baseline*. We assume the baseline is the same as defined in the debit calculation. The implication is that the 50% degree of gain is relative to this condition.
- *Base year*. We assume 2012 is the base year for the analysis (same as the debit calculation), which means the present value multiplier is equal to 1 in that year.

Table 6.1 demonstrates how the per unit credits would be calculated for 1 ha of land that would provide habitat-related benefits for 55 years into the future. The per unit credit in each individual year is equal to the degree of gain in that year multiplied by the present value factor. The present value credits then are summed across the years during which the remedial project generates benefits to calculate the total present value of credits for each unit of remediation (1 ha in this examples) over the lifetime of the remediation project.[5] Thus, the increase in habitat quality services (over the baseline) measured in present value (2012) from the hypothetical

[5] If the benefits would have been provided indefinitely, the present value factor would—after about 100 years—become less than 0.01. In practical terms, this means that benefits occurring 100 years from now and into the future are essentially zero. Thus, we can still estimate a finite per unit credit for remediation projects with perpetual benefits.

Table 6.1 Illustrated example of per unit credit calculations using a non-monetary metric

Year	Unit (ha)	Degree of gain (e.g., % increase in species on site)	Present value multiplier[a]	Per unit credit[b] (DSHaYs)
	(A)	(B)	(C)	(D) = (A) × (B) × (C)
2014	1	10	0.94	0.09
2015	1	20	0.92	0.18
2016	1	30	0.89	0.27
2017	1	40	0.86	0.35
2018	1	50	0.84	0.42
⋮	⋮	⋮	⋮	⋮
2065	1	50	0.21	0.10
2066	1	50	0.20	0.10
2067	1	50	0.20	0.10
2068	1	50	0.19	0.10
Credit per hectare of land remediated				12.08

[a]Present value factor = $1/(1 + \text{discount rate})^{(\text{year}-\text{base year})}$, where discount rate = 3% and base year is 2012

[b]Per unit credit is calculated by multiplying percent gain by present value factor for each unit and for each year of the project

remediation project is equal to 12.08 DSHaYs per hectare of habitat remediation (sum of discounted unit in column (d) of Table 6.1).

Scaling remediation requires that the total debit be divided by the per unit credits. The total debit was estimated in Sect. 4.6.1 to be 319.5 DSHaYs. The per unit credits were estimated above to be 12.08 DSHaYs per unit. Thus, to offset the total loss of 319.5 DSHaYs with the example remediation project, 26.5 ha of remediation would be required, i.e., 319.5 DSHYs/12.08 DSHYs per hectare restored = 26.5 ha.

Therefore, the amount of hectares to be provided each year, that is, remediated this year and then made available for a period of 55 years that will compensate the total interim loss of habitat, is approximately 26.5 ha.

In this example, the remediation required consists of a smaller area (26.5 ha) than the damaged land (100 ha). Although this seems counterintuitive, this is a result of the summation over time and the services provided by habitats over time. In our example, the debit occurs over a period of nine years, with services reduced by 50% for the first five years, and then improves linearly over the final four years back to a baseline level (Table 4.1). As shown in Table 6.1, the credit occurs over a much longer period of 55 years, with services improving linearly over the first four years to 50% in the fifth year, and continuing at a 50% improvement over pre-remediation conditions for 50 years. Although the credit is discounted by 3% each year, the longer time that services are provided by the remediation means that a smaller area is needed to compensate for the debit.

6.3.4 Example: Scaling Remediation Using a Monetary Metric

In our simple VEA example from Chap. 4, we assumed that a popular fishing area was contaminated by a chemical release, which led to a loss of 200 fishing trips per year for a three-year period and a diminished experience for the 100 fishing trips per year that continued at the site for a three year period. With our assumptions, we calculated the total debit to be a discounted lost value (DLV) of €18,938 (see Sect. 4.6.2 and Table 4.2).

Using the value-to-cost approach, scaling would proceed as follows: the Competent Authority would recover the €18,938 from the responsible operator and use that to implement compensatory remedial actions. These actions might include actions such as fish stocking, improving public access to fishing areas, or habitat improvements designed to improve the fishing experience (e.g., improve catch rates, average fish size, or quarry species mixes). Importantly, the amount of remediation would be scaled such that the total cost would not exceed €18,938. In other words, the value-to-cost framework ensures equivalence between the debits and credits by assuming that *the cost of remediation* equals the total debits.

Using the value-to-value approach, the Competent Authority would also recover funds that would be used to implement similar types of remediation. However, the scaled amount of funds used for this remediation would be based on the value the anglers derive from the proposed remediation project, rather than being based on the value of the damage. In other words, the value-to-value framework ensures equivalence between the debits and credits by assuming *the amount of remediation* should be based on the increase in value provided by the remediation project.

To scale the appropriate amount of remedial actions in the value-to-value approach, we follow the methodology described above for non-monetary metrics by estimating the per unit credits and dividing them into the total debits.

Below we identify hypothetical assumptions for this illustrative example, based on the scenario described in Sect. 6.3.3 (Table 6.2 summarises the calculations).

- *Start year*. We assume remediation benefits are first realised in 2014.
- *End year*. We assume benefits will stop being provided in 2068.
- *Unit* (*Table* 6.2, *column A*). We scale the number of fishing trips to the damaged area, that is, unit = fishing trip.
- *Degree of gain in human use* (*Table* 6.2, *column B*). Increased catch rates typically improve the value of recreational fishing. We assume that a proposed remediation project improves the catch rate by 25% to anglers by increasing fish stocks through habitat improvements. We assume this occurs gradually over a five-year period from 2014 to 2018 and then continues to provide that same service gain for the next 50 years, at which point the incremental benefits are no longer achieved.
- *Degree of gain in economic value* (*Table* 6.2, *column C*). To translate this gain in human use into an economic gain (measured by our monetary metric), we make an assumption about the economic value per trip (in real cases, this

Table 6.2 Illustrative example of per unit credit calculations using a monetary metric

Year	Unit (fishing trips)	Degree of gain in human use (% increase in catch rate)	Degree of gain in economic value due to increase in human use (€) (10% of base value of fishing trip (€25))	Present value multiplier[a]	Per fishing trip credit[b] (€)
	(A)	(B)	(C)	(D)	(E) = (A) × (B) × (C) × (D)
2014	1	5	0.50	0.94	0.47
2015	1	10	1.00	0.92	0.92
2016	1	15	1.50	0.89	1.33
2017	1	20	2.00	0.86	1.73
2018	1	25	2.50	0.84	2.09
2019	1	25	2.50	0.81	2.03
2020	1	25	2.50	0.79	1.97
⋮	⋮	⋮	⋮	⋮	⋮
2064	1	25	2.50	0.22	0.54
2065	1	25	2.50	0.21	0.52
2066	1	25	2.50	0.20	0.51
2067	1	25	2.50	0.20	0.49
2068	1	25	2.50	0.19	0.48
Credit (value) per trip from the remediation project					€60.40

Notes [a]Present value factor = $1/(1 + \text{discount rate})^{(\text{year}-\text{base year})}$, where discount rate is 3% and base year is 2012

[b]Per unit credit is calculated by multiplying degree of gain in human use by degree of gain in economic value by present value factor for each unit and for each year of the project. All are expressed per 1 fishing trip

To shorten the table, some of the results were omitted and substituted by the ellipsis

assumption should be based on studies from the literature or primary economic research). We assume that increasing the catch rate by 25% would increase the value of a fishing trip by 10% of the original value of the trip, or €2.50 (current value is €25) per trip. Because this benefit is dependent upon the gain in human use, its trajectory over time mirrors the gradual increase over five years, then becomes constant for the next 50 years.

- *Present value multiplier* (*Table* 6.2, *column D*). We assume a 3% discount rate.
- *Metric*. The monetary metric is the value of the human use of the resource. On the debit side, this was the value of the loss. Here it represents the value of the gain in human use due to the remediation project, that is, the 25% increase in catch rate (€2.50 per trip).
- *Baseline*. We assume the baseline is the same as that defined in the debit calculation. The implication is that the gain in human use (catch rate) is relative to this condition.
- *Base year*. We assume 2012 is the base year for the analysis (same as the debit calculation).

When fully implemented in 2014, this remediation project would improve the value of a recreational fishing trip by €2.50. Using the same implementation schedule as the gain in human use (i.e., gradually reaching the maximum level in five years and then providing constant gains for the next 50 years), Table 6.2 shows the calculations to determine the increase in value of a single fishing trip. Column (E) is the credit per trip and is equal to the number of trips (A) times the degree of gain in human use (B) times the degree of gain in economic value (C) times the present value factor (D). Thus the increase in value (over the baseline of €25) associated with increasing one fishing trip annually due to the remediation project is €60.40, measured in present value (2012) (sum of discounted unit benefits in column (D) of Table 6.2).

We then scale remediation to ensure the value-benefits of the remediation project is equal to the value of the loss, providing us the number of *improved* recreational trips that will offset the loss. Thus, we divide the total debit (€18,938) by the per unit credit (€60.40) and determine that remediation must provide sufficient improvements such that the 25% increase in catch rate is realised on approximately 314 recreational fishing trips annually (€18,938/€60.40 per trip = 314 trips).

In the value-to-value calculations above, the Competent Authority would determine the cost to undertake the required habitat improvements to increase fish stock so that recreational anglers would realise a 25% increase in catch rates. The cost to implement those habitat improvements would then form the basis of the liability claim.

6.4 Estimating Costs of Remediation Options

The total cost of environmental liabilities using an equivalency analysis is equal to the sum of the cost of the efforts to undertake the equivalency analysis, the cost to implement and maintain the required remediation projects, and the costs associated with planning, overseeing, and monitoring the remediation projects. The cost of analysis may include staff costs of the Competent Authority and possibly the costs of hiring external experts (e.g., ecologists, economists, lawyers). Here we focus on the cost of the remediation project because of its importance in comparing different remediation options. In other words, some projects may provide the same level of complementary or compensatory remediation but differ in their costs.

6.4.1 Remediation Cost Components

The results of an equivalency analysis can be presented in terms of the amount and type of required remediation or the cost of implementing the required remediation. Unit costs of the required scale of remediation may include:

- Project design (including scientific or engineering design, permitting, surveying, and other related design costs),
- Project implementation,
- Project administration,
- Operations and maintenance,
- Failure contingency,
- Monitoring and reporting expenditures, and
- Oversight costs by the Competent Authority.

The costs of remediation projects are project specific, but some general considerations on potential cost components are provided in Table 6.3.

6.4.2 Estimating Remediation Costs

Cost estimation requires diligence by those managing the remediation project in order to ensure all cost categories are covered. It is important that scientists and engineers responsible for designing the project provide input to, or at least verify, cost estimates (GHK 2006, eftec 2010).

Cost information typically may be obtained by:

- Developing site-specific remediation costs;
- Acquiring representative costs of similar projects (keeping in mind potential differences related to site location, local economic factors, similarity of resources or projects); and
- Other such factors that may influence variations in project costs or through discussions with experts in ecological remediation and engineering design.

One approach to estimating costs is to rely on actual cost information from previous projects that are similar to the selected remediation alternatives. Cost information can be found in the literature, from documentation of previously conducted projects, or from established cost estimate tables available in some Member States. Important considerations when using the 'cost transfer' approach based on similar projects are (1) to standardise the costs on a per unit or per area basis to control for project size and (2) to ensure characteristics other than size of the documented project(s) are similar to the one under consideration. In addition to project size, other criteria for evaluating similarity might include climate, topography, region (labour and capital costs across regions), time, and other relevant factors.

Note that uncertainty in the cost components of the claim is not addressed in detail in this document. However, the typical approach, which is the addition of a flat-rate contingency to monitoring and oversight costs, is discussed in Sect. 4.2.3 of the United States National Oceanic and Atmospheric Administration Technical paper 99-1 (NOAA 1999). Diekmann and Featherman (1998) also discuss possible ways to assess cost uncertainty.

Table 6.3 Important cost components when estimating remediation cost

Cost	Description
Planning	lanning and design of the remediation project. This may include preliminary ecological (or economic) surveys to identify extent of damage (or loss of value or welfare) and ecological (or economic) surveys to count or assess post-incident ecological data (or loss of value or welfare). This cost component can be subdivided into two parts: • *Initial design, surveying, and plan preparation* covers those aspects of work that are necessary prior to preparing a final executable remediation plan. It should also include the costs of REA. • *Final plan preparation* covers the preparation of a final remediation plan including, as necessary, any public outreach and comment, design drawings, engineering models, survey results, mobilisation schedules, and other required plan elements.
Acquisition of permits	The acquisition of any necessary legal access, permitting requirements, or other such obligations that may be necessary to conduct remediation work
Acquisition of land	Land acquisition costs can cover any necessary costs to acquire property easements, rights-of-use, or other legal instruments needed to implement remediation actions and subsequent operations, monitoring, or adaptive management actions
Implementation	Implementation costs cover the fundamental elements of remediation implementation, including all labour, materials, transport, infrastructure development, site management and oversight, and supplies needed during the implementation process
Operations and maintenance	Operations and maintenance costs cover all costs required to run and manage the project, including necessary labour, equipment, materials, and supplies for these operations. Often this component is expressed as an annual cost of operating and/or maintaining the implemented activity (e.g., annual removal of sediments from constructed drains)
Oversight	Oversight covers any cost associated with necessary oversight of remediation projects by Competent Authorities. This cost component most likely consists of labour costs and administrative overhead costs, that is the additional cost (on top of labour costs) to account for ongoing expense of operating the organisation (rent, communication costs, utilities, permits, insurance, etc.)
Monitoring and reporting	Monitoring and reporting covers all necessary monitoring and reporting costs, including costs of labour, materials, supplies, and information dissemination
Failure contingency	The contingency cost component covers all necessary and appropriate contingency costs that apply to uncertainties associated with remediation project execution. The purpose is to account for unexpected/random events that increase actual costs over planned costs (e.g., bad weather). Often this cost component consists of a standard percentage amount that is added to the best cost estimate (e.g., all costs mentioned above). General practice is to assume an additional 20–40% of total estimated costs as 'contingency costs'

References

Diekmann, J. E., & Featherman, W. D. (1998). Assessing cost uncertainty: Lessons from environmental restoration projects. *Journal of Construction Engineering and Management, 124*(6), 445–451.

Eftec. (2010). *Scoping study on the economic (or Non-Market) valuation issues and the implementation of the water framework directive,* report to European Commission DG Environment, Ref: ENV.D.1/ETU/2009/0102rI.

GHK Consulting LTD in association with RPS Ecology. (2006). *UK biodiversity action plan: Preparing costings for species and habitat action plans costings summary report.* Report to Defra and partners, London.

NOAA. (1999). *Discounting and the treatment of uncertainty in natural resource damage assessment,* Technical paper 99-1, National Oceanic and Atmospheric Administration (NOAA), USA.

Syncera. (2005). *Verkenning argumentatielijnen fasering en doelverlaging (derogaties) Kaderrichtlijn Water.* Rijksinstituut voor Integraal Zoetwaterbeheer en Afvalwaterbehandeling. Arnhem: Syncera.

WATECO. (2003). *Common implementation strategy for the water framework directive (2000/60/ EC), Guidance Document No 1: Economics and the Environment – The Implementation Challenge of the Water Framework Directive.* Working Group 2.6, Brussels.

Chapter 7
Step 5: Monitoring and Reporting

Diana Lane, Kate LeJeune and Joshua Lipton

Abstract Although not a component of equivalency analysis, per se, monitoring and reporting of the progress and efficacy should be an integral aspect of an overall liability assessment. Key steps in the monitoring and reporting phase of a project include remediation planning and implementation, monitoring, and reporting. These steps are described in this chapter.

Keywords Monitoring · Program reporting · Remediation implementation

7.1 Remediation Planning and Implementation

After the equivalency analysis has been performed and remediation projects are selected and scaled (see Chap. 6), a remediation plan is prepared. This plan, which builds on information gathered during the equivalency analysis, should include project goals, implementation details, engineering plans and designs, and biological plans and designs. Development of remediation plans may be iterative and may include (see Fig. 7.1 for simplified steps):

- Initial planning-level design plans;
- Initial site inspection, monitoring, or measurements needed to refine design plans;
- Full implementation plans and design drawings;
- All necessary health and safety plans;

D. Lane (✉) · J. Lipton
Abt Associates, 1881 Ninth St., Suite 201, Boulder, CO 80304, USA
e-mail: Diana_Lane@abtassoc.com

J. Lipton
e-mail: Josh_Lipton@abtassoc.com

K. LeJeune
Le Moulin du Siffré, 66730 Rabouillet, France
e-mail: klejeune@icloud.com

J. Lipton et al. (eds.), *Equivalency Methods for Environmental Liability*,
https://doi.org/10.1007/978-90-481-9812-2_7

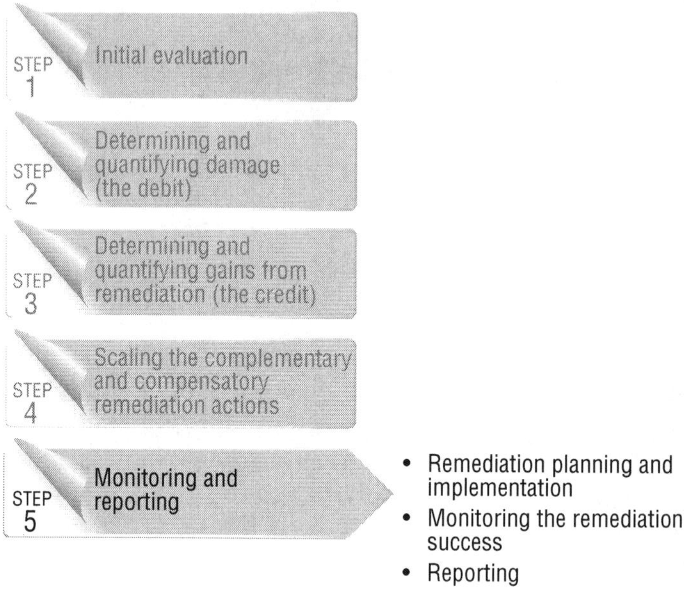

Fig. 7.1 Step 5 of equivalency analysis

- Quantitative performance measures and plans for corrective actions, if needed; and
- Monitoring, adaptive management, and reporting and communication.

In the United States, the usual practice is to first develop a draft remediation plan and then subject this to a review process. Ideally, the process also allows the general public to review the plan and provide suggestions and comments. Following this review process, a final remediation plan should be developed for implementation. Whether such a plan is required and its details will be determined by the official guidance in each Member State. The content of a remediation plan is likely to vary depending on the nature of the remediation actions and the specific phase of remediation. Generally, remediation plans should contain information regarding:

- Intended outcome(s) of the remediation (project goals);
- How the specific activities will contribute to the intended outcomes (including engineering and biological plans and designs);
- The anticipated time period of remediation actions;
- The anticipated duration of ecological recovery;
- The performance standards to be used to gauge project progress and success;
- Monitoring plans;
- Any potential risks to human health, society, culture, or the environment;
- How corrective actions will be taken, if necessary;
- Any necessary ongoing operations and management necessary to ensure project objectives; and

- Project costs, including monitoring, oversight, operations, and management.

Specialised literature, existing general guideline documents, and Member States' requirements should be consulted for additional information on the preparation of remediation plans (e.g., Reinharz and Burlington 1996).

7.2 Monitoring Efficacy of Remediation

Monitoring should be conducted at intervals that are determined based on biological, chemical, physical, social, or economic factors important to the determination of success, including both performance metrics (such as plant cover on revegetated sites) and desired outcomes (e.g., utilization of habitat by birds, changes in fish populations). The metric originally used to quantify debit and credit should still be of value in evaluating project success and benefits achieved, but need not circumscribe all monitoring needs. Typically, monitoring and post-monitoring reporting will be an essential part of remediation plans. Monitoring is undertaken to:

- Gauge the nature, extent, rate, and efficacy of remediation benefits;
- Enable plan modifications, corrective actions, or other adaptive management interventions;
- Protect human health, welfare, or ecological resources from unintended consequences, as well as from predicted near-term risks during stabilisation of remediation actions (e.g., during certain types of remediation projects, such as contaminant dredging, contaminant concentrations may initially increase prior to decreasing);
- Enable more efficient management of natural resources in recovery areas; and
- Contribute to an expanded database of remediation project efficacy and recovery outcomes.

Monitoring also allows assessment of whether implementing parties are doing what they committed to in the remediation plan. Design and performance criteria included in the remediation plans can help Competent Authorities to assess whether responsible parties meet the set requirements during implementation.

Monitoring should be conducted prior to, during, and following implementation of remediation plans. It must be sufficient to quantify remediation gains for the desired assessment metrics that were used to design the remediation plan. However, monitoring may consist of many different types of actions including:

- Chemical monitoring of media (e.g., water, air, soil, sediments) and biota (e.g., fish tissues);
- Biological monitoring of individuals, populations, communities, or habitats;
- Physical and hydrological monitoring of target attributes (e.g., sediment accretion rates, water flows, etc.); and

- Monitoring focused on specific programmatic or performance measures (e.g., aboveground biomass of planted wetland vegetation; contaminant runoff concentrations, hectares placed into conservation easements).

Monitoring plans should be designed to consider a reasonable range of natural variability, including factors such as seasonal variations in hydrographs, wildlife migrations, growing seasons, tidal cycles, and, potentially, human uses. Design of monitoring plans should be statistically based, with appropriate consideration of necessary design elements to discern changes in environmental variables. Finally, all monitoring should be conducted pursuant to scientifically designed and approved sampling and analysis plans. It is important to keep in mind that the costs of monitoring, including reporting, should be incorporated into remediation costs. An example is provided in Box 7.1.

Box 7.1: A framework for Post-remediation Monitoring

Post-remediation monitoring is a key step in the remediation process. An effective post-remediation monitoring plan will help to:

- Identify problems that could be corrected;
- Quantify benefits realised; and
- Provide information that can be communicated to policy makers and the public about the benefits of remediation.

Before developing a post-remediation monitoring plan, the conceptual model for a project must be specified. This model should clearly delineate the remediation action, the expected intermediate outcome, and the pathway/process by which the intermediate outcome will lead to the desired long-term results.

An effective monitoring framework takes advantage of the conceptual model to provide important information for each step of the remediation process. Ideally, the monitoring framework will include both pre-implementation monitoring to determine initial conditions and reference sites that will be monitored simultaneously with the project site. Because baseline conditions can change over time (e.g., a drought may cause a regional decrease in fish populations), monitoring changes in reference conditions over time allows for appropriate adjustments to be made to baseline conditions.

For each step in the monitoring framework, a plan should be developed that specifies who will be responsible for monitoring, to whom results will be reported, the objective of that monitoring step, the monitoring actions to be taken, the location of the monitoring, the timing of the monitoring, and any benchmarks that will trigger corrective action.

Overview of Monitoring Steps

Step 1: Monitor project sites and appropriate reference sites to establish pre-implementation conditions.

Step 2: Monitor implemented action to determine if the implementation has been successful in terms of both implementation criteria and outcomes. These results should trigger corrective actions to implementation if necessary.

Step 3: Monitor project site and reference sites over the short term (often 1 to 5 years) to determine if the implementation has achieved the intended intermediate outcomes. These results should trigger corrective actions if necessary.

Step 4: Monitor project results and reference sites over the long term (often 3 to 10 or more years) to quantify project outcomes.

7.3 Reporting

Case-by-case reporting is not a requirement of the ELD. However, since monitoring and evaluation are the only means by which Competent Authorities can demonstrate that they have protected the public's natural resources, reporting the results of monitoring and evaluation is crucial. Therefore, authorities may wish to consider making damage assessment reports available for public review at regular intervals and in an accessible format.

Monitoring plans also should provide for post-monitoring reporting. Reporting is a critical means of:

- Communicating remediation plan successes (and failures) to the affected public;
- Communicating necessary alterations in monitoring design or anticipated recovery rates to the affected public;
- Communicating any potential human health risks (or lack thereof) to the affected public; and
- Contributing to scientific knowledge regarding remediation efficacy and recovery rates.

Whether publicly available or not, reports should include a description of the project, project goals, the anticipated recovery and benefits trajectory, data collected as part of monitoring, and a synthesis and interpretation of the monitoring data. Any corrective actions taken or anticipated should be reported. The degree of resources and/or services recovery to baseline conditions, and relative to the anticipated recovery trajectory, should be described. In addition, an efficient European Union level reporting mechanism would allow Member States to learn from remediation experiences in other countries. In this context, ensuring reporting of Member States' follow-up on remediation usage and success (e.g., efficacy, recovery outcomes, key parameters) to a central European Union database might be a valuable contribution.

Reference

Reinharz, E., & Burlington, L. (1996). *Restoration planning: Guidance document for natural resource damage assessment under the oil pollution act of 1990, the damage assessment and restoration program* (National Oceanic and Atmospheric Administration Technical Report).

Chapter 8
Economic Valuation for Equivalency Analysis

David Chapman, Scott Cole and Ece Özdemiroğlu

Abstract Economic analysis contributes to the implementation of the Environmental Liability Directive (ELD) through the estimation of changes in the value (in monetary terms) of damaged resources from an incident. As noted in Annex II of the ELD, resource-to-resource and service-to-service approaches should be considered first (paragraph 1.2.2), however, if these methods are not possible or appropriate, then alternative valuation techniques shall be used (paragraph 1.2.3). This chapter provides an overview of the concept of economic value, how it can be estimated, and how it can help in determining compensation for damages to resources and services covered by the ELD.

Keywords Economic value · Economic valuation · Value equivalency analysis

8.1 Introduction

Previous chapters focus on how best to identify, measure, and quantify ecological losses and gains across a wide variety of resources. In some cases, as Annex II of the Environmental Liability Directive (ELD) also acknowledges, ecological equivalency may not be sufficient and value equivalency analysis (VEA) should be used (Paragraph 1.2.3):

D. Chapman (✉)
Rocky Mountain Research Station, United States Forest Service,
240 West Prospect Road, Fort Collins, CO 80526, USA
e-mail: davidchapman@fs.fed.us

S. Cole
EnviroEconomics Sweden Consultancy, Skansstigen, 7B 832 51 Frösön, Sweden
e-mail: scott@eesweden.com

E. Özdemiroğlu
eftec, 4 City Road, London EC1Y 2AA, UK
e-mail: ece@eftec.co.uk

J. Lipton et al. (eds.), *Equivalency Methods for Environmental Liability*,
https://doi.org/10.1007/978-90-481-9812-2_8

> If it is not possible to use the first choice resource-to-resource or service-to-service equivalence approaches, then alternative valuation techniques shall be used. The competent authority may prescribe the method, for example monetary valuation, to determine the extent of the necessary complementary and compensatory remedial measures. If valuation of the lost resources and/or services is practicable, but valuation of the replacement natural resources and/or services cannot be performed within a reasonable time-frame or at a reasonable cost, then the competent authority may choose remedial measures whose cost is equivalent to the estimated monetary value of the lost natural resources and/or services.

The above quotation refers to the so called 'value-to-value' approach (using valuation for estimating both the debit from damage and credit from remediation) and 'value-to-cost' approach (estimating the debit from damage in monetary terms and setting this as the budget to spend on remediation which must be identified using resource or service equivalency). Section 5.2.4 explains when such VEA may be the preferred approach. This Chapter presents the conceptual background to economic value and valuation methods.

The foundations of economic value are in welfare economics, which studies the impact on individuals and society of policies, economic activity, or other changes that may affect human welfare (Johansson 1991). The tools of welfare economics can help assess the following in the context of environmental damage and remediation:

- Identifying a means of measuring human wellbeing (utility, welfare);
- Assessing how a damaged resource changes an individual's wellbeing, compared to a situation without the damage;
- Identifying the amount, type, or quality of a compensatory resource that an individual may be willing to substitute, or trade off, in lieu of a damaged resource;
- Assessing how individuals make temporal and geographical trade-offs between restored resources or services and damaged resources or services;
- Identifying an amount of economic compensation that ensures an individual is no worse off than with a loss;
- Determining how to aggregate individual compensation to societal level and identify the distributional effects of who 'wins' and 'loses' from different compensation schemes; and
- Assessing the implication of different compensation mechanisms.

This Chapter also discusses the use of VEA making a distinction between compensatory/complementary remediation (which requires input from economic analysis) and primary remediation (which does not). A brief discussion of the application of the economic valuation concepts and methods under the Habitats Directive, Wild Birds Directive, and the Water Framework Directive is provided. The chapter ends with a review of how specific application of economic valuation in this context may differ from a more theoretical approach.

8.2 Economic and Environmental Compensation

'Compensation' occurs when an individual is provided something in lieu of his or her first choice; it is based on the notion that an individual's loss of a good can be offset through the provision of an alternative good. Full compensation occurs when the loss is offset without making the individual any worse off than had the loss not happened. The most common method of economic compensation is through the use of money. For example, if a homeowner loses private belongings from a house fire, the insurance company provides economic compensation (minus the deductibles) in the form of a cash payment. The individual is no worse off than before the fire because they can substitute cash as an alternative to their private belongings. The assumption that an individual can make trade-offs that substitute one good for another is critical and is a defining characteristic of economic compensation.

In the context of the ELD, it is assumed that *society* can be compensated for the loss of environmental resources and/or services by the provision of alternative equivalent resources and/or services that are replaced, restored, or rehabilitated. That is, society is willing to substitute restored resources for damaged resources. Substitution implies that some form of measurement can be made across goods: how much of X (credit) is equivalent to how much of Y (debit)? 'Debit' here refers to both to the initial damage and interim loss, as defined in Chap. 2. The question does not apply to primary remediation. Rather, in the ELD context, the question applies to complementary and compensatory remediation.

Three factors need to be considered in answering this question.

The first consideration is the relevant unit of measurement that can be used when comparing changes in 'wellbeing' associated with X to changes in 'wellbeing' associated with Y? This is the economic equivalent of the ecological metrics described in Chap. 4. Economists generally refer to utility as a measurement of individual wellbeing, or welfare. An individual's utility (welfare) generally increases when the individual is able to consume more goods and it decreases when the individual is unable to consume them. The 'goods' here are not limited to market/consumable goods, but also include environmental quality, cultural factors that influence wellbeing and so on.

In the context of the ELD, the only loses and gains that are of concern are social welfare losses, which are the decline in utility (wellbeing) from not being able to enjoy a resource, such as a nature reserve, or a service, such as fishing or hunting. Loss of profits that may accrue to a tour operator or other business that relies on a resource or service is excluded from this discussion of (environmental) compensation in the context of the ELD.

Second, we need to address *how much* of X is needed to compensate for the loss of Y, given our utility unit of measurement. This is conceptually equivalent to the habitat scalars discussed in Chap. 5. Depending on the specifics of the individual incident, we can choose from a number of different welfare measures (see the Annex to this Chapter, Sect. 8.7), which determine how individuals are impacted by different incidents. In determining how much compensation is necessary, three specific issues, similar to ecological equivalency, need to be addressed: (1) the type(s) of resources and/or services damaged and those that are provided as complementary and compensatory remediation; (2) geographical location of damage; and (3) remediation and time periods over which damage is experienced and remediation is provided. Chaps. 4–6 provide a fuller list of factors that need to be considered.

Finally, we need to address the compensation mechanism, that is, goods being provided in lieu of an individual's first choice of no damage to (loss of) environmental goods and services from the incident. Two mechanisms that we will focus on include (1) an amount of money and (2) an amount of resources or service provision that would provide adequate remediation, i.e. would make an individual no worse off between the damaged resource and the remediated resource. Although economic theory supports both, the ELD excludes the first mechanism of monetary compensation to individuals in lieu of lost resources or services. Thus, although we may measure environmental loss or gain using monetary and/or non-monetary metrics, the provision of compensation must be resource based.

Economic theory also provides insight into how resources are substituted (compensated) over time through the use of discounting. Discounting acknowledges that people have preferences over when events occur. Typically, gains are preferred sooner rather than later, and losses are preferred later rather than sooner. The degree of these preferences are called the rate of time preference and commonly referred to as the discount rate in equivalency analysis. Discounting can also be applied when the non-monetary (resource) units are used. The details of discounting are discussed in Chap. 4 (see Box 4.4).

8.3 Definition of Economic Value

Economic values, in the context of compensation under the ELD, are the values placed by individuals on environmental resources and their services. Economic values are expressed in relative terms based on individuals' preferences for given changes in the quality and/or quantity of resources and services. Their preferences, in turn, are determined by how the changes in the resource or service affect their wellbeing (or utility or welfare). The theoretical basis for this is discussed in the Annex to this chapter (Sect. 8.7).

Preferences are measured by how much individuals are willing to trade off between alternative goods that affect their wellbeing—including money. Money is used as a common unit as it is a familiar metric to all people, it is divisible, and it makes comparison of financial and environmental costs and benefits possible. The choice of money is not a comment about the 'value of or value for money'. Using this unit, preferences are measured in terms of individuals' willingness to pay (WTP) money to avoid an environmental loss or to secure a gain, and their willingness to accept (WTA) money as compensation to tolerate an environmental loss or to forgo a gain.

As mentioned above, what is estimated by economic valuation is the value of a marginal change. The intention is never about estimating the (absolute) value of the environment, but rather the change in environmental conditions brought about due to the incident. The specific use of WTP or WTA is based on the presumed property rights. Should individuals and/or society have the right to an undamaged environment, then the theoretically correct measure of welfare changes is WTA—what is the minimum amount of compensation necessary to allow the environmental change to occur. If not, the theoretically correct measure is WTP—what is the maximum amount that individuals or society would be willing to pay to not have the change occur. In practice, WTP is the most commonly applied measure, and it typically considered to be a lower bound estimate of WTA (Mitchell and Carson 1989).

Individuals have several motivations for having positive WTP and WTA to protect ecosystem services. These motivations are analysed within the so called Total Economic Value (TEV) typology (Fig. 8.1). The 'total' here refers to the sum of different motivations rather than the absolute value. For a breakdown of ecosystem services that provide welfare to individuals and hence for which they have economic values, see Chap. 2 (Box 2.3). The motivations for preferences, or the types of value, can be summarised as follows:

Use values involve some interaction with the resource, either directly or indirectly:

- *Direct use value*: Environmental goods and ecosystem services are used in either a consumptive manner, such as industrial water abstraction, or in a non-consumptive manner such as for recreation (e.g. hiking).
- *Indirect use value*: The value of ecosystem services provided such as nutrient cycling, habitat provision, climate regulation, water regulation and assimilative capacity, soil formation, carbon sequestration, air filtration etc.
- *Option value*: Not associated with the current use of ecosystem services, but rather the benefit of keeping open the option to do so in the future. A related concept is *quasi-option value* which arises through avoiding or delaying irreversible decisions, where technological and knowledge improvements can alter the optimal management of an ecosystem.

Non-use values are associated with benefits derived from resources and services unrelated to their use. Non-use value can be split into three parts:

- *Altruistic value*: Derived from knowing that contemporaries can enjoy ecosystem services;
- *Bequest value*: Associated with the knowledge that ecosystems and their services will be passed on to future generations, and
- *Existence value*: Derived simply from the satisfaction of knowing that ecosystems continue to exist, regardless of use made of it by oneself or others, now or in the future.

Those who make direct and indirect use of ecosystem services, i.e. the users, are likely to hold both use and non-use values. Those who do not directly or indirectly

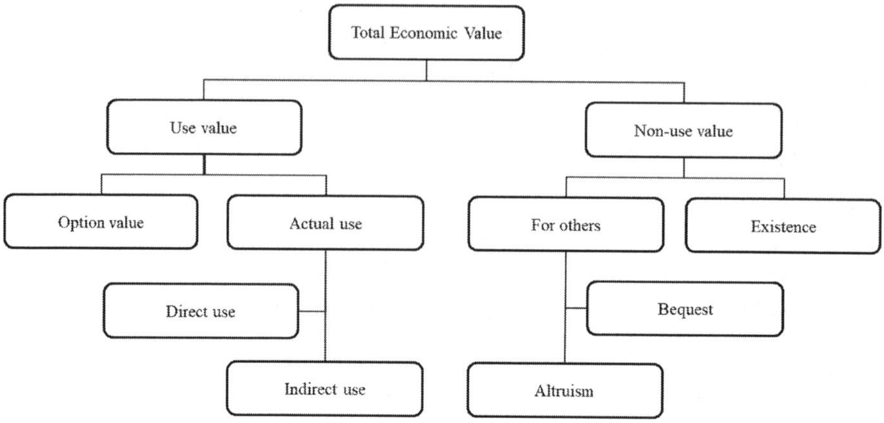

Fig. 8.1 Total Economic Value

use a service, but still hold non-use values, are called non-users. While users are relatively easy to identify, there is no theoretical definition of non-users. The size of the relevant non-user population depends on factors such as the uniqueness of the resource/service affected and whether the change is negative and irreversible. In practice, the definition is an empirical question which can be answered by primary research. In the context of the ELD, welfare losses (increases) experienced by both users and non-users of environmental resources and services should be included in debit and credit estimates.

8.4 Measuring Economic Value

Economic valuation methods have been specifically designed to estimate economic value in monetary terms. They all follow three key steps: qualitative and quantitative assessment of the change and monetary valuation, as shown in Fig. 8.2. Therefore, even though it is 'economic' value that is being estimated, it would not be possible without the multidisciplinary input through ecological or other scientific/technical analyses that estimate the impact (debit from loss or credit from remediation).

The economic valuation methods differ in terms of what kind of data they use and which value components they cover:

Market prices and consumer behaviour data measure the actual direct spending (e.g. visitor fee, recreational expenses, spending on bottled water and other defensive expenditure) and can be used to estimate the values held by the users of a resource.

Many goods and services provided by natural resources, such as ecosystem services, are potentially market goods (e.g. provisioning services (food, drink,

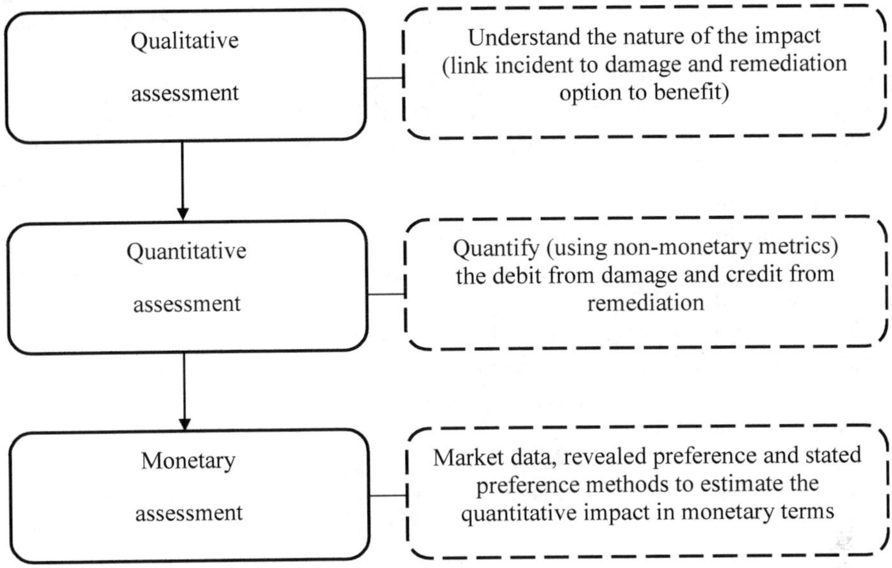

Fig. 8.2 Three key steps of economic valuation

fibre), tourism etc.). The market price at which a good is exchanged reveals some, but incomplete, information about its economic value. In particular, the price paid in a market transaction reveals the amount of money the buyer is willing to give up to obtain the good and the amount of money the seller is willing to accept as compensation for giving up the good. Thus, for example, the economic value of commercial fishing (a provisioning service) is estimated at the market value of the fish catch. Similarly, financial revenues from tourism (a cultural service) reflect the market value of this service.

Market price information, however, is an imprecise indicator of the economic value of a particular ecosystem service, since it does not fully reflect true WTP or WTA. For example, many buyers may be willing to pay more than the market price to obtain the good. The difference between the maximum amount a buyer is willing to pay and the actual price paid is termed consumer surplus, reflecting the element of benefit from obtaining the good that is a surplus to the consumer. Similarly, the seller of the good may be willing to accept a lower amount than the market price to give up the good. The difference between the minimum amount a seller is willing to accept and the actual price received is termed producer surplus, reflecting the additional benefit in exchange gained (in effect 'economic profit'). Overall, in the case of market goods and services, economic value (WTP or WTA) is reflected by the market price paid or received, plus any consumer or producer surplus.

There are also many ecosystem services (in fact the majority of services) that are not traded in markets, and are consequently 'un-priced' or 'non-market' goods. To estimate the economic value of these non-market goods and services in the absence of price information, two types of valuation methods are developed:

Revealed preference methods involve analysing the consumption behaviour of individuals in markets that are related to the resources or services to be valued. Choices consumers make in these markets can be used to reveal the value places on non-marketed goods and services. One such method is called 'travel cost' which estimates a value that a user implicitly places on a site (heritage, recreation, etc.) from the travel costs they incur getting to the site and back, including the cost of time spent in the journey and at the site. Such costs include any costs of transportation whether in private car or public transport, accommodation, spending on food, drink, and recreational activities. The other main revealed preference method is 'hedonic pricing', which analyses property sales data to estimate the premium various characteristics of a property fetches in the market. Such characteristics can include the type and size of the property but can also be location related attributes such as cleanliness of the area, beauty of the landscape, nearness to cultural sites, schools, transport systems etc.

Revealed preference methods can only estimate the value of the changes that have already been experienced, and are limited to the values held by users. For the 2010 Gulf Oil Spill, the United States government undertook a travel cost study to estimate the loss of recreational use of the gulf as a result of the spill (Tourangeau et al. 2017) (See Box 8.1).

Box 8.1: Gulf of Mexico Oil Spill—Recreational Lost Use Study

On April 20, 2010, the Deepwater Horizon oil rig exploded and caught fire in the Gulf of Mexico, resulting in an oil spill that released more than 130 million gallons of oil into the waters of the Gulf. It was one of the largest oil spills in the history of the United States, much larger than the 11-million- gallon *Exxon Valdez* disaster in 1989. Oil came ashore at beaches, marinas and marshland in Texas, Louisiana, Mississippi, Alabama, and western Florida and affected the marine and coastal ecology as well as recreational beach use, fishing, and boating in many parts of these areas. This box summarises the economic analysis of the recreational beach use impacts. Other analyses aimed at assessing other damage components are not covered here.

In the United States, under the Oil Pollution Act (OPA) of 1990, NOAA and other federal and state agencies (collectively known as the 'Natural Resource Trustees') are responsible for assessing the damages. To estimate recreational use impacts from the oil spill, the Trustees undertook three travel cost studies specifically designed to estimate the lost recreational use and associated lost public value in an area that measured more than 16,000 km of shoreline. Together the studies represented perhaps the largest undertaking of recreational use studies, involving the analysis of nearly half a million aerial photographs, 35,000 onsite counts of beach visitors, and 129,000 interviews over a period of three years. Multiple studies were implemented to address differences in use patterns, opportunities to substitute to other locations, and the values associated different types of use. The studies were designed to focus on specific categories of Gulf Coast use; beach users from within the

Gulf Coast states; beach users from outside of the Gulf Coast states; and people who recreated using boats. The studies were designed to measure the baseline level of use in each of the categories, potential for substitution to other locations; and collect information to estimate the value of recreation trips by beach users from local areas, from beyond the local areas and boaters.

Based on the results of the studies, it was estimated that there were about 16 million fewer beach, boating and fishing visits because of the spill. The total combined lost recreational use value was estimated at approximately €619 million (2015 prices[1]). For additional detail see: https://www.doi.gov/deepwaterhorizon/adminrecord, Sect. 5.10.

Sources: Tourangeau et al. (2017) and Deepwater Horizon Natural Resource Damage Assessment Trustees (2016).

Stated preference methods involve asking individuals to trade off changes to the resource or service with money. Respondents' stated responses to choices poised to them can be used to reveal the value they place on the changes in non-marketed goods and services. The responses are either in terms of individuals' WTP to avoid a loss or to ensure an improvement or WTA to tolerate a loss or to forgo an improvement. There are two main variations to stated preference methods which vary in the way the trade-off question is designed: (1) contingent valuation, which asks WTP or WTA for changes in a combined set of resources or services and (2) choice experiments, which can evaluate changes in resources or services in terms of their characteristics (or attributes) and asks respondents to choose options with more or less of one or more attributes. The case study in Chap. 11 uses the stated preference method to estimate both the debit and the credit from the forest fire in Catalunya. The Exxon Valdez case in the United States also used a contingent valuation design to estimate the economic value of the loss suffered by users and non-users due to an oil spill in Alaska (Carson et al. 1992). This case was an important milestone in getting stated preference studies accepted as legal evidence in the damage assessment cases in the United States and encouraged the publication of the best practice guidance by the National Oceanographic and Atmospheric Administration (NOAA) (Arrow et al. 1993). For a more up to date guidance on both contingent valuation and choice experiment approaches, readers could consult Bateman et al. (2002) or Johnston et al. (2017). For the 2010 Gulf Oil Spill in the United States, the governments implemented a national level contingent valuation study to measure the total value of the loss to users and non-users of the Gulf of Mexico. (Bishop et al. 2017) (See Box 8.2).

[1]Using 30 June 2015 inter-bank spot exchange rate of United States $1.12 to €1.

Box 8.2: Gulf of Mexico Oil Spill—Contingent Valuation Study

On April 20, 2010, the Deepwater Horizon oil rig exploded and caught fire in the Gulf of Mexico, resulting in an oil spill that released more than 130 million gallons of oil into the waters of the Gulf. It was one of the largest oil spills in the history of the United States, much larger than the 11 million gallon *Exxon Valdez* disaster in 1989. Oil killed or impaired numerous plants and animals in the open ocean, ocean floor, and along more than 1,600 km of shoreline from Texas to Florida. Oiling at beaches, marinas and marshland affected recreational beach use, fishing, and boating in many parts of these areas. Oil coated hundreds of kilometers of marsh habitats, killed fish, sea turtles, dolphins, birds and both shallow and deep water corals. Sixteen million fewer recreation trips occurred as a result of the spill (see Box 8.1). The ecological effects of the spill are anticipated to last for many years.

In the United States, under the Oil Pollution Act (OPA) of 1990, NOAA and other federal and state agencies (collectively known as the 'Natural Resource Trustees') are responsible for assessing the damages. To estimate damages from the oil spill, the Trustees undertook a total value study using the contingent valuation method.

The study developed and extensively tested a survey instrument designed to measure the value the United States public places on preventing another incident similar to the Gulf Oil spill. This is one measure of the lost total value the public places on such an incident. The final surveys interviewed a random sample of more than 2000 American adults who were told about (1) the state of the Gulf before the 2010 accident ('baseline' in the ELD terminology); (2) what caused the accident; (3) damages to Gulf natural resources due to the spill and people's uses of those resources; (4) a proposed program for preventing a similar accident in the future; and (5) how much their household would pay in extra taxes if the program mentioned in (4) were implemented.

To develop the survey instrument, survey designers worked closely with natural scientists studying the damages, and the recreational loss team, to understand the types and degree of damages that occurred and the timeframes until resources and uses would return to baseline conditions to ensure accuracy of the information provided to survey respondents. The survey designers also worked closely with oil drilling and policy experts to understand options to prevent a similar spill in the future. All of the collected information was pre-tested multiple times to ensure understandability by survey respondents. Multiple 'show cards' were developed to present the information to respondents, provide pictures of the resources impacted, and show the potential technological changes that could be implemented to ensure that a similar spill did not happen in the future. The final surveys were administered in-person by experienced and trained professional interviewers using computer and the physical show cards.

Based on results of the study, the estimated lost total value (both use and non-use values) the United States population placed on the Gulf Oil spill was €15.4 billion (2015 prices[2]). For additional detail see: https://www.doi.gov/deepwaterhorizon/adminrecord, Sect. 5.15.3.

Source: Bishop et al. (2017).

The economic valuation literature is large and growing. This is evidenced by the number of studies in the largest of the online databases, the Environmental Valuation Reference Inventory[3] (EVRI). EVRI is searchable on a number of variables including where the study was located, the focus of the study (general environmental assets, the type of environmental goods and services valued, the environmental stressor or the source of the stressor), the valuation technique, available study information (e.g. questionnaires, maps, tables, etc.), economic measures, and the year of the study.

Databases like EVRI are particularly useful in facilitating the subsequent use of economic value evidence from literature especially when primary economic research is not possible. The process of using previous evidence is known as **value (or benefits) transfer**. Unit value estimates or functions of factors that explain the variation in economic values (WTP or WTA) could be transferred from an existing study to a new analysis context. While there are limitations to value transfer (in particular in finding the evidence from the literature that is appropriate to the analysis in hand), it has advantages, in terms of less time and funds needed compared to undertaking a primary valuation study. Good value transfer is crucial in expanding the use of combined ecosystem services and economic value approach (See eftec (2010) for the best practice guidance in the United Kingdom). The case study in Chap. 12 uses value transfer to estimate the debit and credit in the context of the impacts of water abstraction from a river.

Choosing the appropriate valuation method depends on a number of factors including which value type is to be estimated, the specifics of the incident, what data are available and/or can be collected and time and resource availability. These should be assessed as part of the initial evaluation in Step 1 of the equivalency analysis (Chap. 3). The assessment of available data and the scale of damage (and hence the appropriate level of analytical effort) will also help with decide whether value transfer is sufficient or whether primary research would be required. Once it is established that economic valuation (and hence VEA) is required (see Sect. 5.2.4 for when this could be the case), the following criteria can be used to select the appropriate methods:

- Use (or user) values can be estimated using market prices, revealed and stated preference methods.
- Non-use values can only be estimated using stated preference methods.

[2]Using 30 June 2015 inter-bank spot exchange rate of United States $1.12 to €1.

[3]www.evri.ca.

- While individual non-use values tend to be smaller than individual use values, non-user population tends to be much higher. Therefore, in the aggregate, non-use values could be a substantial sum and should not be ignored as a potential category of value.
- If the resource or service affected has no market or is not linked to a marketed good or service (e.g. a free access recreation site), there would not be any market price data and revealed and stated preference methods should be considered.
- There are many resource/service/incident specific factors such as the location, time and duration of damage, or whether baseline can be reached or not that are relevant for qualitative and quantitative assessment of debits and credits and play a part in selecting the appropriate valuation method.

8.5 Environmental Compensation in Practice

In this section we discuss how the theoretical economic framework applies under an equivalency analysis approach. One of the key foundational issues is the legal requirement that the compensation mechanism be in the form of units of resources remediated, rehabilitated, or replaced (rather than money). We discuss these differences below. Section 8.6 addresses the resulting policy implications.

Resource-based compensation required by the ELD. Although compensation may be *measured* (or scaled) in monetary or non-monetary units, compensation can only be provided in resource-based units. Thus, the equivalency analysis approach to compensation is differentiated from the economic welfare approach by the requirement for a resource-based compensation mechanism, rather than money. This leads to two consequences. First, a 'pay-back' of what was lost through the provision of primary remediation (*R*) in Fig. 8.3 will not be sufficient because of interim loss (where there is a time delay for resources and service to recovery to their baseline conditions) and payment of compensation does not occur immediately, but rather is delayed either for assessment or legal reasons. Both of these delays require that the amount of compensation due to the public needs to be increased to off-set the losses and make the public whole. This explains the need for compensatory and complementary remediation in addition to primary remediation (*R*). Second, compensation is 'paid' to society as a whole through remediation of a resource that, by definition, is a non-divisible public good (see below).

Compensation on a society (not individual) level. When evaluating the impacts of a *public* action such as compensation, one economic approach is to consider social indifference curves that identify the trade-offs in welfare gains and/or losses by different individuals or segments in society. There is no universally accepted method for making comparisons between individuals' utility changes (or their

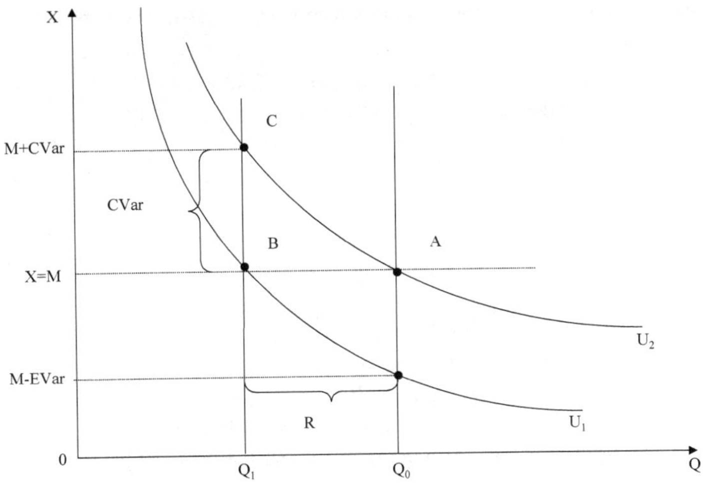

Fig. 8.3 Compensating an individual for a reduction in the non-market (environmental) good Q. (*Note* The graph represents the simplified case where the public good is free, i.e., q = 0). *Source* A reconstruction of Fig. 3–10 in Freeman (2003b)

heterogeneous preferences[4]) when considering whether to proceed with a public action. The theoretical compensation mechanism (e.g., Hicks-Kaldor compensation principle[5]) outlines proceeding with a public project (compensation) if it is at least theoretically possible for those who benefit from such a project could compensate those who lose in such a manner that no one is made worse off. Under the theoretical approach to compensation that relies on money as the compensation mechanism, different amounts of money could be paid to different individuals to ensure compensation satisfies every individual (Flores and Thacher 2002). In the case of equivalency analysis, the social trade-off is between an amount of damaged resources and an amount of remediated resources, where the latter is a non-divisible compensation mechanism. Thus, compensation can only satisfy society in the aggregate; it is impossible to satisfy specific individuals with heterogeneous preferences when the compensation mechanism is a non-divisible resource (Jones and Peace 1997).

Utility versus resource units or services. In the theoretical economic framework, the amount of compensation for an environmental loss is based on individuals' utility change due to the interim loss, that is, comparison of utility *before the incident* to *after incident* (which is then aggregated across individuals). This requires that we

[4]By heterogeneous preferences we mean that individuals have different reactions to loss and gain depending on many factors such as type and frequency of use, socio-economic characteristics and personal preferences.

[5]Kaldor-Hicks compensation principle states that at 'least in theory' gainers can compensate losers, and both would still be better off than without the proposed change.

first estimate the shape of a person's indifference curve (their preferences) and then sum the discounted utility change each year over the time period of the damage. Equivalency analysis relaxes the theoretical assumption that compensation be based on an individual's change in utility and, instead, handles it in one of two ways. First, in the case of a non-monetary metric, compensation is based on the change in the resource or service units (e.g., number of hectares, number of birds, litres of water) rather than interpreting how that change affects individual wellbeing. Second, when equivalency analysis uses a monetary metric, the loss (and also the gain in the case of value-to-value approach) is based on the individual's change in utility, which is translated into money through economic valuation.

Table 8.1 summarises some of the differences between the theoretical approach and the equivalency analysis approach to environmental compensation. Although the table mentions 'units' in the case of habitat and resource equivalency analyses, the damage and benefits of compensation can be estimated as a whole (e.g., VEA, which uses economic valuation approaches to estimate the WTP for the entirety of a compensation package rather than the benefits provided by a unit of compensation).

8.6 Implications of the Resource Equivalency Approach

Some of the policy implications of the slightly modified equivalency analysis approach to compensation are discussed below which describe the extent to which society's losses are offset through resource-based compensation. Note that most, but not all, of the modifications apply to the case of non-monetary metrics.

Constant value across individuals (homogenous preferences). Equivalency analysis does not account for heterogeneous preferences when providing compensation (this applies under both monetary and non-monetary metrics). This approach, which is driven by the legal requirement to provide a non-divisible compensation mechanism (resources), provides the amount of resources and services based on the aggregate across those impacted by the incident. At the level of a specific individual, the amount of resource compensation provided may be more than they need or not enough is assumed to make this substitution, even if all individuals may not agree to it.

Constant value across quantities (non-diminishing returns). In theory, a non-monetary metric that measures environmental loss and gain assumes a simple linear relationship between the physical environmental change (e.g., acres lost/gained, number of fish lost/gained, gallons of water lost/gained) and the value society places on that change. For example, if contamination of a 10-ha wetland reduces the services to zero on five of the hectares, a HEA or REA approach (e.g., discounted service hectare years) assumes that half the value society places on that particular wetland would be lost. This may be a good approximation, but it may not. For example, if those 10 ha were the last 10 ha of wetland available in this region, society may place a very high value on it and the loss in value may be greater than 50%. If the assumption that the scale of value loss is equal to the scale of the

Table 8.1 Theoretical versus equivalency analysis approaches to environmental compensation

Compensation considerations	Theoretical economic welfare approach	Equivalency analysis (monetary metric)	Equivalency analysis (non-monetary metric)
Compensation mechanism	Money, private goods, or environmental goods	Legal requirement that compensation be resource or service based	
Compensation on the individual versus society level	Aggregates (sums) individual utility changes then, if money is the compensation mechanism, assumes that 'winners' could theoretically compensate 'losers' (Hicks-Kaldor), ensuring compensation satisfies each individual in society	Does not aggregate individual utility changes; instead, relies on a compensation mechanism that is non-divisible, which means that compensation can only satisfy society in the aggregate and not on the individual level (where preferences are assumed to vary across people)	
Metric for measuring damage	Estimates a utility function for a resource/service, then measures utility change resulting from environmental loss and gain; utility change can be translated to monetary or non-monetary metric	Assumes a utility function and measures change in utility (translated to money) resulting from damage (measurement of gain varies depending on value-to-value or value-to-cost approach)	Measures unit[1] change in a non-monetary metric and assumes it approximately reflects changes in wellbeing
Compensation for un-substitutable or irreplaceable resources	Without the substitutability assumption, economic compensation is not possible; such losses result in 'infinite' economic compensation	Equivalency analysis framework allows for compensation by distinguishing between irreplaceable and un-substitutable. An un-substitutable resource cannot be compensated for, but an irreplaceable resource can be replaced by a resource that is 'of a similar type or quality'.	

[1]A unit represents either a resource (e.g., number of fish or birds) or service (e.g., provision of habitat). *Note* that the damage and benefits of compensation can be estimated as a whole, too

physical damage is not appropriate, then a non-monetary metric may not be appropriate[6] or adjustments to this assumption should be made.

In summary, the use of a non-monetary metric implies that the existence of a species, or a habitat acre, is equally valued across quantities and, as described above, across individuals (see also values over time, below). This type of problem —where a metric for loss and gain fails to reflect nonlinear changes in value—does not arise in the case of a monetary metric because that approach measures an individual's change in utility. The *change in utility* approach generally accounts for the difference in the scale of the damage through the shape of the utility function, for example, a birdwatcher's utility function will reflect an extremely high loss of value when 90% of a bird population is lost due to contamination but may reflect a much lower loss of value when only 5% of a population is impacted. In contrast, a simple application of the *change in the resource* approach (a non-monetary metric) assumes that the marginal benefit of an additional resource unit is constant (independent of scope). Again, adjustments to these base assumptions can be made to account for specifics of the impact on the species affected.

Constant values over time. The basic equivalency analysis framework holds the value of a resource constant over time (no discounting) and assumes that a replaced resource has equal value to the lost resource (this applies under both monetary and non-monetary metrics). This approach was motivated by the presumed difficulty in estimating the economic value of damaged and remediated resources (Unsworth and Bishop 1994). This inherent assumption may not hold if remediated hectares provided in the future are scarcer than the hectares lost today. Importantly, this constant value over time is not inconsistent with our use of a positive time preference, which (all else equal) assumes that we are impatient and place higher value on benefits that occur today and less value if we have to wait for them (see below).

Discounting resource units, not utility. While discounting utility changes (in monetary units) is a common practice in economic analysis, in resource and habitat equivalency, resource and service units are discounted. Other than the difference in units, the same principles apply and are discussed in Chap. 4 (in particular Box 4.4).

An assumed proxy for the level of environmental resource or service damage. The equivalency analysis approach rests on the assumption that a non-monetary metric is a good proxy for describing the environmental change that occurs (both credit and debit). For example, if we count the number of lost salmon in a polluted river, we assume that this 'indicator species' is a good measure of all the (complex) environmental changes occurring as a result of contamination. Similarly, we assume that quantitative gains in the salmon population will bring with it the ecological benefits that were lost as a result of the damage. As noted above, we not only assume that this metric captures the environmental changes accurately but that these

---

[6]An alternative might be to quantify the loss such that losses higher than a selected threshold are compensated to a greater extent (a scalar).

changes approximately mirror the changes in value the public holds for these resources.

8.7 Annex

8.7.1 Economic Theory of Compensation

In this Annex we discuss the economic theory that underlies the concept of economic compensation. It provides a basis for understanding how economics can inform adequate amounts of environmental compensation.[7]

Economic compensation is based on the notion that an individual's loss of a good or environmental resource can be offset through the provision of alternative goods or alternative environmental resource or services without making the individual any worse off than before the loss. For economic analysis, the most common choice for measuring the extent of a loss is money. However, through the assumption of substitutability, compensation can be measured equally well in terms of the quantity of a good (i.e., the substitution of a damaged resource for a restored resource).[8]

The discussion in this Annex shows how the microeconomic theories of welfare provide a theoretically correct approach for measuring, or scaling, compensation in terms of a monetary metric or a non-monetary metric. It also discusses how the use of resource equivalency requires a number of simplifying assumptions (modifications), leading to a slightly less rigorous welfare approach to compensation.

8.7.2 Indifference Curves

Because economics is primarily the study of how goods and services are allocated in society, economists often begin with a simple starting point: individuals wish to consume (demand) certain goods that firms are willing to produce (supply) at different prices. The focus in this section is on the individual, rather than on the firms. The basic assumption underlying most economic studies is that individuals are rational. Among other things, this means they choose to consume goods that maximise their wellbeing, or utility.

To keep it simple, we will assume individuals maximise their utility by demanding (1) private goods and services purchased in the market and (2) environmental goods and services that cannot be purchased in a market (e.g., a clean

[7]This discussion is based on text from Freeman (2003a) and Flores (2003) published in Champ et al. (2017).

[8]Note that unlike money, compensatory resources may only be able to be provided in distinct units and not be completely divisible. This is not discussed further but should be noted.

environment, bird populations). We are interested in estimating the size of the utility loss when an individual cannot consume an environmental good due to environmental damage. Theoretically, the individual utility loss (aggregated to a societal level) informs the size and extent of the necessary offsetting compensation.

In order to study an individual's choice between private and environmental goods, we need to measure individual preferences for the type and amounts of each. To do this, we assume individuals prefer a certain bundle of private goods, which we refer to as *X*, and non-market environmental goods, referred to as *Q*. It is assumed that an individual's preference for each of these two goods can be described using an indifference curve, as shown in Fig. 8.4. An individual is indifferent between the amounts of the two goods anywhere along the indifference curve U_1, but prefers indifference curve U_2 over U_1. Higher indifference curves represent higher levels of utility to an individual.

The key assumption is that individuals can substitute across goods and still maintain the same level of wellbeing, or utility. It is assumed that if the quantity of the environmental good, *Q*, is decreased, then it can be compensated for by an increase in the quantity of the private good, *X* (or, equivalently, money, which can purchase the private good) and that the individual is entirely indifferent between these alternative combinations of goods (the inverse is also true).

Note also that the curves slowly approach each axis without ever touching the axes, indicating that substitution possibilities are always available. An obvious shortcoming of the substitutability assumption is that in some cases there may not a reasonable substitute: for example, a unique environmental resource and/or irreversible change.

We can illustrate this lack of substitutability through an alternatively shaped indifference curve. Figure 8.5 illustrates an indifference curve for an individual who is not willing to substitute indefinitely between a damaged environmental resource, Q_2, and a substitute remediated environmental resource, Q1. That is, he/she is willing to accept the loss of Q_2 by receiving more Q_1, but only up until Q_2^0. In other words, adequate compensation—in the form of more Q_1 is not possible if Q_2 drops below Q_2^0. Substitution is no longer acceptable. We may interpret Q_2^0, for example, as the minimum viable population of some species.

8.7.3 Compensating Individuals: Measures of Gain and Loss

Given an understanding of how welfare economics measures wellbeing (changes in utility) and how it measures preferences for alternative bundles of goods (indifference curves), we turn to identifying measures and mechanisms for compensating individuals for a loss. We summarize this approach by referring back to the three issues introduced in Sect. 8.2.

First, how should a negative impact on wellbeing be measured? All environmental damage directly impacts (decreases) an individual's utility level. As noted

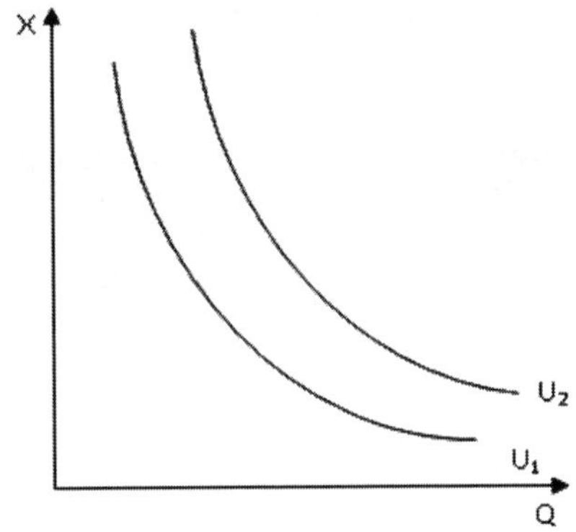

Fig. 8.4 Social indifference curves in the equivalency analysis framework

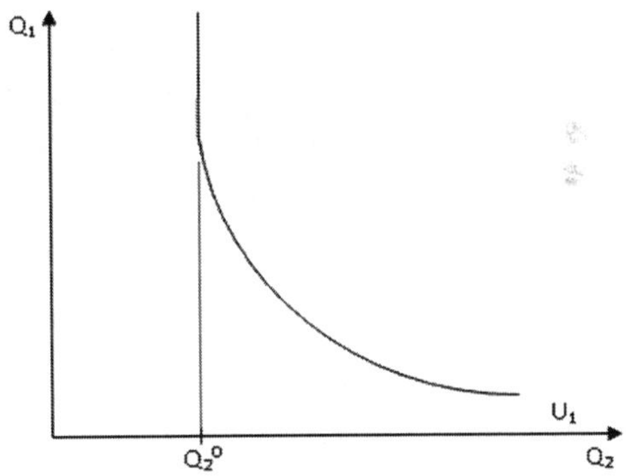

Fig. 8.5 Individual indifference curves without substitution paste Q_2^0

above, utility is a measure of an individual's inherent wellbeing, or welfare, and it is this change in utility that provides the theoretically correct measurement to assess compensation.

Second, what wellbeing measure can be used to compensate for this negative impact? A common measure of how exogenous changes impact individuals is consumer surplus, or the total benefit of consuming the good (utility gained) minus the cost (utility lost) to pay for it. Consumer surplus is useful in measuring the change in utility that may arise from a change in price or quantity of a 'priced' good. In the current context, we are interested in cases where the quantity or quality of a non-priced environmental resource is reduced.

Instead, we want to define a welfare measure in terms of the underlying (observable) indirect utility function (i.e., what factors determine an individual's utility level). Our measure assumes that the initial utility level is constant (U^0) and associated with the status quo, that is, the right to an uncontaminated environment (Q^0). Our measure must answer the following question: given a reduction in Q, what is the appropriate amount of compensation for the individual to be indifferent between the initial clean environment and a contaminated one? The answer,

Compensating Variation (CVar),[9] is the theoretically correct welfare measure, which is also relevant in the context of the ELD.[10]

Asking this question is similar to asking an individual to accept an undesirable change (e.g., contaminated wetland) in exchange for some compensation payment. Therefore, it is also referred to as a person's WTA compensation for a deterioration of the environment.[11] In principle, an individual should be willing to accept an amount equivalent to the CVar in order to be returned to his or her level of utility without the incident.[12] Figure 8.3 demonstrates this concept. Assume an individual is at point A on indifference curve U_2 (based on his or her demand for goods and the supply of such goods by firms) and then the quantity of the public good decreases due to damage from Q_0 to Q_1. For a given amount of a private good ($x = M$, indicating that all income, M, is spent on the private good), this causes the individual to fall from U_2 to U_1 (point B). The compensating amount (in terms of private goods) that would be required to make this individual indifferent (on the original utility curve) between the original and damaged environment is the CVar (i.e., the person is indifferent between point A and point C because they both provide the same level of utility).

Finally, what is an appropriate compensation metric for measuring environmental damage (loss)? In Fig. 8.3, CVar is represented as a vertical distance on the Y-axis and represents a quantity of private goods (X). Therefore, one appropriate metric that compensates for environmental damage and ensures an individual is 'no worse off' than before the damage is to provide an amount of private goods (X), or money that can buy private goods, equivalent to CVar. The size of CVar is estimated through various types of economic valuation, where individuals are asked to state the amount of money (a monetary metric) they would require as compensation to ensure they remain on their original utility level.

In addition to these two monetary metrics (private goods or money), a third way of measuring compensation for environmental damage is to measure the horizontal distance on the X-axis, represented by R (remediation). This resource-based metric is also referred to as a non-monetary metric. Therefore, another equally appropriate metric to compensate for environmental damage that ensures an individual is no worse off than before the damage is to provide an amount of public goods (Q) that is at least equivalent to R. The size of R can be estimated by asking individuals to

---

[9]Some economic textbooks may refer to this as compensating surplus, which is the same thing.

[10]An alternative wellbeing measure might be *Equivalent Variation* (EVar). However, it assumes that the new utility level associated with Q_1 (the damaged environment) is the basis of comparison. Because we assume that the public has an inherent right to the level of utility associated with the undamaged environment (Q_0), we rely instead on CVar.

[11]If, instead, the hypothetical question asked an individual to pay for a desired environmental improvement before it happened, it would be referred to as a WTP, which is associated with the CVar welfare measure (see footnote above). The reason is that CVar is defined relative to the initial level of utility. The improvement increases utility and the payment equalizes utilities in the two states.

[12]Note that the concept of *substitutability* is crucial to this argument, see above.

state the amount of resources (a non-monetary metric) they would require as compensation to ensure they remain on their original utility level.

There are two important caveats to Fig. 8.3. First, the figure depicts a scenario that assumes that the damaged environment is *immediately* 'given back' to the individual. If the environment takes time to recover (and we assume a positive time preference), the amount *R* is insufficient compensation because it does not account for the interim loss (Jones and Pease 1997). More specifically, *R* represents primary remediation by improving the quality and/or quantity of the damaged resource itself. In this sense it is better described as an attempt to 'give back what was lost' to the individual rather than compensation for an interim loss. Thus, compensatory (complementary) restoration is over and above *R*. This is described further below. Second, Fig. 8.3 identifies different *wellbeing measures* for determining 'how much is enough' individual compensation, but the actual *provision* of compensation must be resource based (i.e., money payments are prohibited) as discussed in Sect. 8.5.

References

Arrow, K., Solow, R., Portney, P.R., Leamer, E.E., Radner, R & Schuman, H. (1993). *Report of the NOAA Panel on Contingent Valuation.*

Bateman, I. J., Carson, R. T., Day, B., Hanemann, M., Hanley, N., Hett, T., et al. (2002). *Economic Valuation with Stated Preference Techniques: A Manual.* London: Edward Elgar.

Bishop, R., Boyle, K., Carson, R., Chapman, D., Hanemann, W. M., Kanninen, B., et al. (2017). Science Policy Forum. *356*(6335).

Carson, R. T., Mitchell, R. C., Hanemann, W. M., Kopp, R. J., Presser, S. & Ruud, P. A. (1992). A Contingent Valuation Study of Lost Passive Use Values Resulting From the Exxon Valdez Oil Spill, report to the Attorney General of the State of Alaska.

Champ, P. A., Boyle, K. & Brown, T. C. (2017). *A Primer on Nonmarket Valuation Second Edition*, Kluwer Academic Publishers, the Netherlands: Kluwer Academic Publishers.

Deepwater Horizon Natural Resource Damage Assessment Trustees (2016), *Deepwater Horizon Oil Spill: Final Programmatic Damage Assessment and Restoration Plan and Final Programmatic Environmental Impact Statement.* Retrieved from http://www.gulfspillrestoration.noaa.gov/restoration-planning/gulf-plan.

eftec (2010). *Valuing Environmental Impacts: Practical Guidelines for the Use of Value Transfer in Policy and Project Appraisal.* Report to the UK Department for Environment, Food and Rural Affairs, London.

Flores, N. E., & Thacher, J. (2002). Money, who needs it? Natural resource damage assessment. *Contemporary Economic Policy, 20*(2), 171–178.

Flores, N. E. (2003) Conceptual Framework for Nonmarket Valuation. In Champ, P. A., Boyle, K. and Brown, T. C. (2003). *A Primer on Nonmarket Valuation*, Netherlands: Kluwer Academic Publishers.

Freeman, A. M., III. (2003a). Economic valuation: What and Why. In Champ, P. A., Boyle, K. and Brown, T. C. (Eds.), *A Primer on Nonmarket Valuation.* Netherlands: Kluwer Academic Publishers.

Freeman, A. M., III. (2003b). *Measurement of Environmental and Resource Values.* Washington DC, USA: Resources for the Future.

Johansson, P. O. (1991). *An Introduction to Modern Welfare Economics.* Cambridge: Cambridge University Press.

Johnston, R. J, Boyle, K. J, Adamowicz, W., Bennett, J., Brouwer, R., Cameron, T. A., et al. (2017) Contemporary Guidance for Stated Preference Studies, *Journal of Association of Environmental and Resource Economists*, *4*(2).

Jones, C. A., & Pease, K. A. (1997). Restoration-based compensation measures in natural resource liability statutes. *Contemporary Economic Policy, 15*(4), 111–122.

Mitchell, R. C., & Carson, R. T. (1989). Using surveys to value public goods: The contingent valuation method. *Resources for the Future*. New York, London.

Tourangeau, R., English, E., McConnell, K., Chapman, D., Flores Cervantes, I., Horsch, E., et al. (2017). The Gulf Recreation Study: Assessing Lost Recreational Trips from the 2010 Gulf Oil Spill. *Journal of Survey Statistics and Methodology, 5,* 281–309.

Unsworth, R. E., & Bishop, R. C. (1994). Assessing natural resource damages using environmental annuities. *Ecological Economics, 11*(1), 35–41.

Part III
Case Studies

Chapter 9
The Vistula River Crossing in Poland

Joshua Lipton, Zenon Tederko, Eric English and Magdalena Kiejzik-Głowińska

Abstract This case study examines the environmental damage caused by the construction, between 1998 and 2000, of the Yamal Western Europe gas pipeline where it crosses the Vistula River in central Poland. Because damages occurred in the past, an *ex post* perspective is used to quantify the environmental damage. This case study is provided for illustrative purposes only. The Environmental Liability Directive does not apply retrospectively to any damage caused before 30 April 2007. The Yamal Gas Pipeline damaged protected habitats during its two year construction period. Terrestrial damages were caused by excavation and associated disturbance. Replanting following construction limited interim losses. Aquatic damages were caused by excavation of the riverbed, deposition of excavated dredge spoils into the channel, and sedimentation from the excavation and deposition processes. No primary remediation action was performed for the aquatic environment. Equivalency analysis in this case study was used to calculate the debits (damages) both for a primary impact zone caused by excavation, and a secondary impact zone associated with disturbance. The use of habitat scalars enabled us to normalise damages to index habitat types. Compensatory remediation projects were selected to benefit riparian forest and shallow-water habitats. For riparian forest remediation, two alternatives were deemed feasible: replanting and preservation. The preservation alternative generated slightly more benefits than the replanting

J. Lipton (✉)
Abt Associates, 1881 Ninth Street, Suite 201, Boulder, CO 80302, USA
e-mail: Josh_Lipton@abtassoc.com

Z. Tederko
Pro-Biodiversity Business Services Ltd, Al. Komisji Edukacji Narodowej 46/1, 02-797 Warsaw, Poland
e-mail: zenon.tederko@gmail.com

E. English
Bear Peak Economics, 623 Maxwell Avenue, Boulder, CO 80304, USA
e-mail: Eric.English@bearpeakeconomics.com

M. Kiejzik-Głowińska
Eko-Perspektywa, ul. Kołłątaja 7/29, 80-262 Gdańsk, Poland
e-mail: mkg@ekoperspektywa.pl

J. Lipton et al. (eds.), *Equivalency Methods for Environmental Liability*,
https://doi.org/10.1007/978-90-481-9812-2_9

alternative, but benefits were highly sensitive to the estimated annual probability of development of the forest to be preserved. For aquatic remediation, in-stream channel improvements were scaled as the preferred alternative.

Keywords Habitat scalar · Gas pipeline · Terrestrial damage · Aquatic damage Poland

9.1 Introduction

This case study presents a conceptual evaluation of environmental damages caused by the construction of the Yamal gas pipeline across the Vistula River in central Poland. The study focuses on calculating environmental damages to terrestrial and aquatic habitats within both an excavation area in which the pipeline was laid (the primary impacts zone) and a wider area more indirectly impacted by construction (the secondary impact zone).

This case study is organised as follows. First, we describe the incident and include a discussion of the affected locations, habitats, and affected services. We then determine and quantify damage (debits) and analyse benefits (credits) gained through compensatory remediation actions. Finally, we scale compensatory remediation actions and discuss uncertainties in the analysis.

It should be emphasized that the Environmental Liability Directive (ELD) does not apply retrospectively to any damage caused before 30 April 2007 (or where the incident that caused the damage took place and ended before that date). Therefore, the damage described in this case study that was caused between 1998 and 2000 is not within the scope of the ELD.

It also should be noted that the ELD imposes strict liability, i.e. irrespective of negligence or fault, to 'operators'[1] of activities listed in Annex III of the ELD in relation to damage to water, protected species and habitats, and land.[2] For other activities, it imposes only fault-based liability, i.e. if negligence or fault can be proven, and only in relation to damage to protected species and habitats. Pipeline projects are not listed in Annex III. Therefore, ELD would likely apply only if the resulting damage to protected species and habitats is deemed significant and if the developers are found to have been at fault or negligent.

Finally, we note that the ELD requires primary remediation measures to be carried out without delay and at the site where damage happens. In this case, mitigation actions were taken following pipeline construction. We treat these actions as the effective analogue of primary remediation in order to determine the residual damage for which compensatory remediation should be provided.

---

[1] See Article 2(6), ELD.

[2] Article 3(1)(a) and (b), ELD.

Section 9.2 describes the project and the affected habitat and services. Section 9.3 quantifies debits from environmental damage, while credits from remediation are quantified in Sect. 9.4. Scaling of remediation to sufficiently offset damages is discussed in Sect. 9.5. The chapter ends with a brief discussion of uncertainties and conclusions (Sects. 9.6 and 9.7, respectively).

9.2 Description of the Incident

A pipeline crossing of the Vistula River caused substantial impacts to habitats and biological resources in an area of high natural and landscape value. The area in which the pipeline was constructed contains a number of protected species and habitats, including:

- Candidate[3] Special Areas of Conservation[4] to be designated due to the occurrence of natural habitats that are listed in Annex I of the Habitats Directive (HD) (e.g., marshy willow meadows, old riverbeds and small water bodies, and others);
- Protected fauna species listed in the Annex II of the HD;
- A Special Protection Area[5] listed for its breeding and migratory bird species (under Appendix I of the Wild Birds Directive); and
- The Ciechocinek Lowlands, a Protected Landscape that was created to protect the local microclimate and the landscapes bordering the Vistula.

The transit gas pipeline crossing of the Vistula Valley was constructed between 1998 and 2000. The following actions were performed:

- Construction of a building site on the west bank of the Vistula River on agricultural land. Two sections of the gas pipeline, each 1.2 km in length, were prepared at this site, which has a surface area of 12 ha (1,200 × 100 m). The building site did not cause significant environmental damage within the scope of the ELD because it was located entirely on agricultural land.
- Completion of pressure tests on the prepared sections of the pipeline. This required withdrawal of 1,800 m^3 of river water, followed by the release of the water at a rate of 3.0 m^3/min for 10 h. Water that was subjected to pressure tests was not polluted but was deoxygenated. However, because of the large flow of the Vistula, the uptake and release of this water was deemed not to have posed a significant threat to the aquatic environment (the average flow in the Vistula is 930 m^3/s, hundreds of times greater than the flow of the deoxygenated water).

[3]At the time of preparing this case study, the Natura 2000 Network was in the process of being established in Poland.

[4]Of the PLH040004 habitats of the Włocławek Vistula Valley.

[5]PLB040003 Lower Vistula Valley.

- Excavation of the river bed and floodplain. The main river bed was excavated to a depth of 6 m and a width of 100 m; the floodplain was excavated to a depth of 3 m and a width of 100 m. These excavations required the displacement of 1.3 million m^3 of soil, of which about 0.2 million m^3 came from the river bed.
- Following the installation of the gas pipeline across the Vistula, the excavation area was refilled with approximately 0.6 million m^3 of fill.

Construction of the pipeline crossing involved a number of different stressors that damaged aquatic and terrestrial habitats. Habitat stressors included the direct destruction of aquatic and terrestrial habitat through excavation activities. In addition, secondary stressors included noise and vibration caused by equipment, which are known to disturb both birds and fish. Finally, the construction phase included several types of sediment impacts, including placement of dredge spoils from the excavation area into the river bed upstream of the pipeline crossing, as well as sedimentation of areas downstream of both the excavation zone and the dredge spoil location.

Following construction, additional sedimentation effects, as well as sonic disturbance from pipeline operations, may have continued to negatively impact aquatic habitats. These post-construction effects were not considered in this case study because of a paucity of data.

9.2.1 Affected Environments

Downstream from the town of Włocławek, the Vistula flows through a primary channel, as well as a series of braided sandbars and small islands. The floodplain contains oxbow lakes and several small peat bogs. The river banks and riparian corridor are characterised by a mosaic of willow shrubs, riparian/wetland forests, and arable land and pastures. In some places, the river is bordered by high escarpments overgrown by xerothermic swards and dry forests. The east bank of the river, which contains a series of small islands, sandbars, and riparian forests, provides important habitats for birds and water mammals. The west bank consists of steep floodplain terraces covered by grasses, willow shrubs, and xerothermic swards.

The river crossing was constructed within the Important Bird Area of International Importance E-39, which covers about 120 km of the Vistula Valley from Włocławek to the river mouth. Conservation targets at this site include breeding habitats for endangered bird species, overwintering habitats for waterfowl, and habitat for migratory birds.

Downstream of Włocławek, the Vistula historically supported a rich assemblage of fish species. Dam construction in Włocławek and water pollution have substantially reduced this aquatic biodiversity. As a result, the Vistula was classified as a Heavily Modified Water Body under the Water Framework Directive. Nevertheless, it was classified as a Natura 2000 site.

Habitats damaged from pipeline construction included both terrestrial and aquatic habitats, including riparian/wetland forests (Appendix I, HD, code: 91E0a), riparian shrub communities (Appendix I, HD, code: 6430), and sandbar/grassland habitats (not listed in the HD but very important for birds). Damaged aquatic habitats included deepwater habitats of the main channel and shallow-water habitats associated with the shoreline, islands, and oxbows. The agricultural/grassland habitats on the west bank were not considered to be significantly damaged because they provide relatively little ecological value.

9.2.2 *Primary Remediation Undertaken*

A series of mitigation actions were undertaken following construction. We treat these actions as primary remediation for purposes of our case study. Primary remediation actions included recontouring and replanting river banks affected by excavation. These actions were largely successful in restoring terrestrial vegetation. In addition to the revegetation actions undertaken to restore affected terrestrial areas, fish stocking was also performed as mitigation for aquatic impacts. More than 200,000 juvenile fish (including pike, catfish, trout, pikeperch, orfe, salmon, and salmon-trout) were stocked in the Vistula following construction.

9.2.3 *Potentially Affected Services*

The construction period, which lasted almost two years, likely affected a variety of ecological and economic services in the Vistula Valley, including causing the temporary loss of:

- Avian breeding habitats;
- Avian feeding sites;
- Resting places for migratory birds;
- Fish habitats;
- Angling possibilities (through physical impediments to angling, as well direct impacts to fisheries); and
- Other recreational uses.

For this case study, we focus on the loss of ecological services. As noted below, any loss of social, economic, or recreation-related services was determined to be modest and is likely to have been at least partly offset by primary and compensatory remediation of ecological services.

9.2.4 *Potential Social, Economic, and Transboundary Issues*

The surrounding areas provide a mix of residential, industrial, and agricultural land uses. The area does not support any significant tourism. Construction of the pipeline crossing was unlikely to have caused significant social, economic, or transboundary impacts.

9.3 Quantifying Debits from Environmental Damages

Habitat Equivalency Analysis was employed to quantify damages to affected habitats in the Vistula Valley. Value Equivalency Analysis was not considered because in-kind remediation projects are feasible and because no unique economic or social values are believed to have been lost that would not be fully restored through habitat remediation.

Equivalency analyses were performed separately for terrestrial and aquatic habitats. Terrestrial habitats included were riparian/wetland forest, riparian shrub, and sandbar/grassland. Aquatic habitats considered were deepwater, main-channel and shallow-water habitats characteristic of the river bank, oxbows, and island areas.

9.3.1 *Baseline Conditions*

Baseline refers to the quantity and quality of the affected resources and services had the damage causing incident not happened. In this case study, baseline conditions consisted of functional terrestrial and aquatic habitat types. Because the disturbance involved complete destruction of extant habitat and because remediation actions are designed to be undertaken in the Vistula watershed, a quantitative description of pre-impact baseline conditions is not necessary. Rather, we describe damages relative to the habitat conditions characteristic of the project area.

9.3.2 *Quantification of Damages*

For both terrestrial and aquatic habitat types, losses were quantified within primary and secondary impact areas. The primary impact zone is the excavated area for the terrestrial habitats and the dredge-spoil placement area for the aquatic habitats. The secondary impact zone for terrestrial habitats is the disturbance region extending 50 m on either side of the excavation site. This secondary impact zone is partially damaged as a result of physical trampling of undergrowth, noise and light disturbance, and other anthropogenic factors. The secondary impact zone for aquatic habitats consists of habitat damaged by sedimentation from the excavation and from the dredge spoil area.

9.3.3 *Terrestrial Habitats*

The degree of loss of ecological functions in the primary impact zone was determined to be 100%. This is based on the complete removal of all habitats during excavation. A total of 2.4 ha of riparian forest, 3.1 ha of riparian shrub, and 2.6 ha of sandbar/grassland were damaged in the primary impact zone. Cumulative ecological service loss in the secondary impacts zone was estimated by local biologists to be 30% in the riparian forest habitat, 15% in the riparian shrub habitat, and 15% in the sandbar/grassland habitat based on the relative degrees of disturbance. A total of 1.6 ha of riparian forest, 2.0 ha of riparian shrub, and 3.0 ha of sandbar/grassland were estimated to be damaged in the secondary impact zone.

Immediately after the construction period, primary remediation was undertaken in terrestrial areas. Recovery times for replanted terrestrial areas were estimated based on published studies and personal communications with local biologists. Recovery of the riparian forest habitat was assumed to take 40 years; shrub habitat recovery was assumed to be 10 years. Recovery of the sandbar/grassland habitat was assumed to be complete within one year. Recovery rates, estimated based on consultation with local ecologists, were assumed to progress linearly for all three habitat types.

Table 9.1 summarises these ecological damage assumptions for the terrestrial habitat types. Figure 9.1 illustrates the loss trajectories for both primary and secondary impacts zones.

Table 9.1 Vistula River Crossing—ecological service loss assumptions for terrestrial primary and secondary impact zones

Habitat type	Damaged area (ha)		Ecological service loss (%)		Duration of loss during construction (months)	Recovery period (years)	
	Primary	Secondary	Primary	Secondary		Primary impact zone	Secondary impact zone
Riparian forest	2.4	1.6	100	30	20	40	1
Riparian shrubs	3.1	2.0	100	15	20	10	1
Sandbar/ grassland	2.6	3.0	100	15	20	1	1

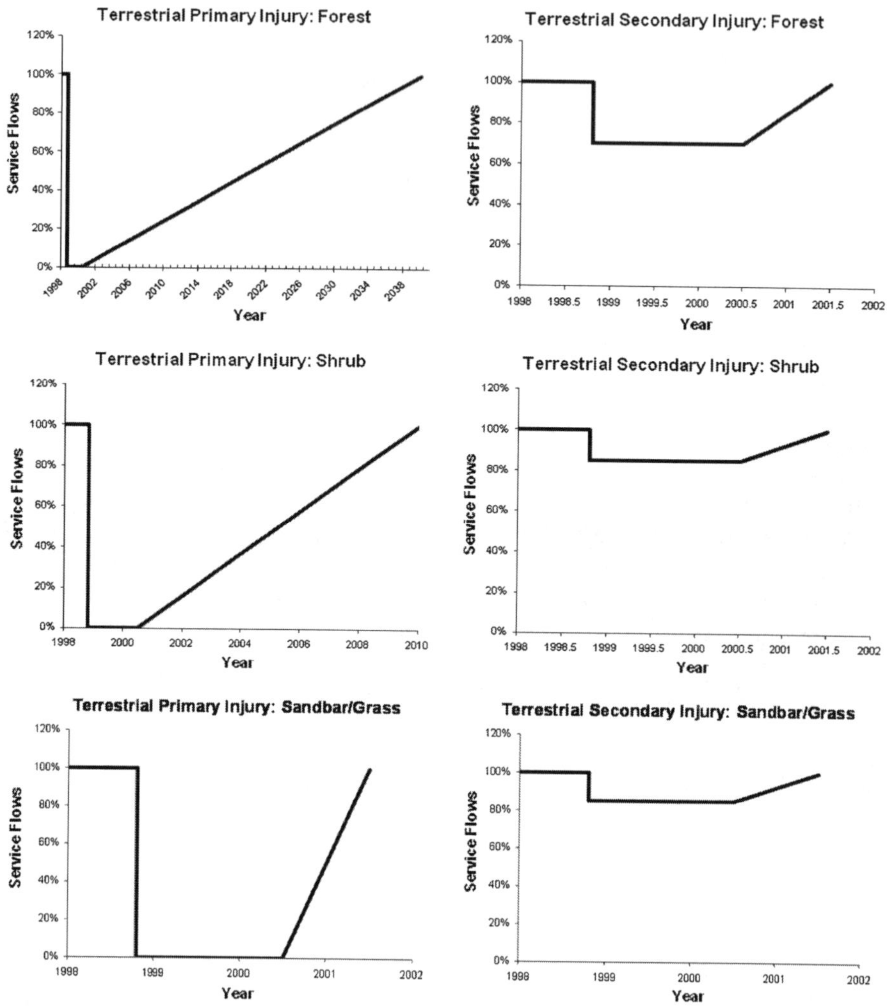

Fig. 9.1 Illustration of percent ecological service loss and recovery within primary and secondary impact zones for riparian forest habitat (top two panels), riparian shrub habitat (center panels), and sandbar/grassland habitat (bottom panels)

9.3.4 *Aquatic Habitats*

As with the terrestrial habitat types, ecological function losses within the primary impact zone were assumed to be 100% to reflect the effects of the excavation process. A total of 6 ha of deepwater, main-channel habitat and 4.4 ha of shallow-water habitat were damaged by the excavation. In addition, 10 ha of deepwater, main-channel habitat was lost from deposition of the dredge spoil.

The secondary impact zone damaged by sedimentation effects included a reach of the Vistula downstream of the excavation site, as well as the reach downstream of the dredge deposition site. Visual observations of the river bottom by local biologists indicated that adverse effects of sedimentation extended 1 km from the primary impacts zone. To quantify the degree of damage, we applied an exponential decay function to this 1-km river reach. The exponential model was fit to assign 100% loss to the upstream edge of the 1-km secondary impact zone, declining to approximately 0% loss at the downstream edge of the secondary impact zone (Fig. 9.1). Integration of the loss over this exponential decline region yielded an average effective loss of cumulative ecological service functions of 18.8% over the 1-km distance. A total of 40 ha of deepwater habitat was damaged within each of the two secondary impact zones (downstream of the excavation area and downstream of the dredge spoil area). A total of 8.4 ha of shallow-water habitat was damaged within the secondary impact zone downstream of the excavation area.

Recovery rates for the aquatic habitats were estimated by local aquatic biologists to be three years. Table 9.2 presents the various damage assumptions for the aquatic habitat types. Figure 9.2 illustrates the loss and recovery assumptions used to calculate debits.

9.3.5 *Calculation of Debits*

Total debits associated with the habitat-level damages are calculated using the information presented above. We used 2007 as the base year of our calculations, with the excavation impacts occurring, as noted above, between 1998 and 2000. We used a discount rate of 3% to calculate the present value of Discounted Service Hectare Years (DSHaYs). Table 9.3 presents an example of the debit calculations for the riparian forest primary impact zone, which results in 36.47 DSHaYs (rounded up to 37 for ease of presentation in the next tables). Tables 9.4 and 9.5 summarise the results of the debit calculations for all terrestrial and aquatic resources, respectively.

Table 9.2 Vistula River Crossing—ecological service loss assumptions for aquatic primary and secondary impact zones

Habitat type	Damaged area (ha)		Ecological service loss (%)		Duration of loss during construction (months)	Recovery period (years)	
	Primary	Secondary	Primary	Secondary		Primary	Secondary
Excavation area							
Deep	6	40	100	18.8	20	3	3
Shallow	4.4	8.4	100	18.8	20	3	3
Dredge soil area							
Deep	10	40	100	18.8	20	3	3
Shallow	–	–	–	–	20	3	3

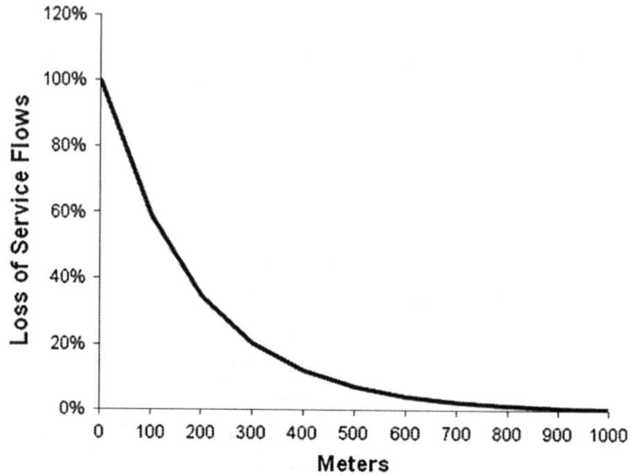

Fig. 9.2 Secondary impacts from sedimentation: exponential relationship between ecological service loss and distance downstream from primary impact zone

Table 9.3 Vistula River Crossing—Habitat Equivalency Analysis calculations for 2.4 ha of riparian forest in primary impact zone

Year	Annual ecological service loss (%)	Total annual service loss (ha)	Discount factor	Total annual discounted service loss (DSHaYs)
	(A)	(B) = 2.4 × (A)	(C)	(D) = (B) × (C)
1998	100.0	2.40	0.99	2.36
1999	100.0	2.40	0.96	2.31
2000	98.8	2.37	0.93	2.21
2001	96.3	2.31	0.91	2.09
2002	93.8	2.25	0.88	1.98
2003	91.3	2.19	0.85	1.87
2004	88.8	2.13	0.83	1.77
2005	86.3	2.07	0.81	1.67
2006	83.8	2.01	0.78	1.57
⋮	⋮	⋮	⋮	⋮
2038	3.7	0.09	0.30	0.03
2039	1.2	0.03	0.29	0.01
2040	0.0	0.00	0.29	0.00
Total				36.47

Note To shorten the table, some of the results were omitted and substituted by the ellipsis

A total of 65 DSHaYs of debit were calculated for terrestrial habitats, with approximately 60% of the debit occurring in the riparian forest habitat type. A total of 123 DSHaYs of debit were calculated for aquatic habitats, with more than 80% of the debit occurring in the deepwater/main-channel habitat.

Table 9.4 Vistula River Crossing—total DSHaYs of debit calculated for terrestrial habitats

Habitat	Impact zone		Total
	Primary	Secondary	
Riparian forest	37	1	38
Riparian shrubs	19	0.7	20
Sandbar/grassland	6	1	7

Note The figure for primary impact zone damage for riparian forest is rounded from Table 9.3

Table 9.5 Vistula River Crossing—total DSHaYs of debit calculated for aquatic habitats

Habitat	Primary impact zone		Secondary impact zone		Total
	Excavation area	Dredge spoil area	Excavation area	Dredge spoil area	
Deepwater	20	33	25	25	103
Shallow water	15	–	5	–	20

9.4 Quantifying Credits from Remediation

In this section we present 'habitat scalars' that were used to normalise remediation alternatives to a single, preferred habitat type. We then quantify the environmental benefits, or credits, of conducting compensatory remediation actions. Finally, we calculate the total hectares of compensatory remediation required to compensate for the debit.

9.4.1 Terrestrial Habitats

For terrestrial habitats, discussions with local biologists indicated that riparian forest habitats are relatively scarce, preferred habitats for a number of bird and mammal species, and represent a more desirable target for habitat remediation than either shrub or sandbar/grassland habitats. To reflect the greater ecological value of the riparian habitat type, a weighting index, or habitat scalar, was used to normalise damage to the forest habitat type. Based on relative scarcity, natural values, and biodiversity indices, the weighting factors selected by local biologists were:

1 ha riparian forest habitat	=	10 ha riparian shrub habitat	=	50 ha sandbar/grassland habitat

Table 9.6 Vistula River Crossing—terrestrial debits normalised to riparian forest habitat using habitat scalars

Habitat type	Unadjusted DSHaYs	Habitat scalar	Forest-normalised debit (DSHaYs)
	(A)	(B)	(C) = (A)/(B)
Riparian forest	38	1	38
Shrub	20	10	2
Sandbar/ grassland	7	50	0.1
Total			40

Table 9.6 demonstrates the normalisation of the debit calculations shown above to riparian forest habitat. As shown in the table, the 65 DSHaYs of terrestrial debit equate to 40 DSHaYs of forest-equivalent habitat damage.

9.4.2 *Aquatic Habitats*

For aquatic habitats, a similar approach was used to reflect the relative ecological value of shallow-water habitats compared to the deepwater/main-channel habitat type. Discussions with local fisheries biologists and ecologists indicated that a habitat scalar of 10:1 would reflect the greater productivity and diversity of the shallow habitats. Table 9.7 demonstrates the normalisation of the aquatic debit to shallow-water habitat so that 123 DSHaYs of unadjusted deepwater habitat loss equates to 30 DSHaYs of shallow-water-equivalent damage.

9.4.3 *Identification of Compensatory Remediation Options*

Two remediation options were identified for terrestrial habitats (both focused on the high-priority riparian forest habitat type):

1. Replanting of agricultural/fallow areas to create new forest, and
2. Acquisition and preservation of existing riparian forest to protect it from future development.

Table 9.7 Vistula River Crossing—aquatic debits normalised to shallow-water habitat using habitat scalars

Habitat type	Unadjusted DSHaYs	Habitat scalar	Deepwater/main channel-normalised debit (DSHaYs)
	(A)	(B)	(C) = (A)/(B)
Deepwater/ main-channel	103	10	10
Shallow water	20	1	20
Total			30

Both projects were deemed to be desirable and feasible by local experts.

For aquatic habitats, several remediation options were considered, including

- Remediation of fish spawning habitats;
- Reintroduction of fish species through stocking;
- Angling management;
- Channel habitat improvements, including creation of side channels (oxbows, shallows), creation of islands, creation of in-stream pools, and addition of wood or large rocks to enhance habitat;
- Navigation controls to reduce sediment scouring caused by vessel traffic;
- Construction of vegetation buffer zones to reduce sediment inputs from agricultural fields; and
- Development of aquaculture infrastructure.

We evaluated these options against the criteria presented in Chap. 5 which were relevant to this case study (Table 9.8). Based on this evaluation process, we determined that channel improvements represented the preferred remediation alternative.

9.5 Calculating Project Benefits and Scaling Compensatory Remediation

In order to calculate project benefits, we evaluated the benefits expected to accrue following project implementation and the recovery rates for each project, as described below.

9.5.1 Terrestrial Remediation

For the forest replanting option, we assumed a 40-year recovery period, consistent with the recovery period used in the debit calculations. We also assumed that forest replanting would generate a net ecological service gain of 90%. This value was selected based on the assumption that the habitat to be replanted currently provides habitat functions roughly equivalent to that of shrub habitat (which has a 1:10 habitat scalar relative to mature riparian forest). The increase in habitat services provided by forest replanting is illustrated in Fig. 9.3.

For the forest preservation option, we assumed that the forest currently generates 100% of habitat services. The added benefit of this option is to sustain this level of ecological function by protecting the habitat from future development. Thus, if development is prevented, a net service gain of up to 100% might be achieved. However, such gains would only occur if there is a real threat of development to be avoided. To calculate benefits, it is therefore necessary to predict the likelihood of development. For example, given a probability of development of 5% in the first

Table 9.8 Vistula River Crossing—evaluation of potential aquatic remediation projects

	Remediation project type						
Evaluation criterion	Fish spawning habitat	Fish reintroduction	Angling management	Channel improvements	Navigation control	Vegetation buffers	Aquaculture
Initial screening							
Nexus to damaged resources	+	+	+	+	+	+	+
Legal	+	+	+	+	+	+	+
No risks to public	+	+	+	+	+	+	+
Feasible	+	+	+	+	+	+	+
Don't cause harm	+	+	+	+	+	+	+
Public acceptance	+	+	?	+	?	+	+
Detailed screening							
Address priority resources	+	+	+	+	+	+	?
Use reliable methods	+	+	+	+	+	+	+
Costs < benefits	+	+	+	+	?	+	?
Readily scaled to debits	Low	Low	Low	High	Low	Low	Low
Benefits can be quantified	High	Low	Low	High	Low	High	High
Consistent with regional planning	High	High	Low	High	Low	High	High
Provides benefits to multiple resources	High	Low	High	High	High	High	Low
Enhance public enjoyment	High	High	Low	High	Low	High	High
Provide long-term benefits	High	High	?	High	?	High	Low

Note We applied either +/− or high/low rankings to habitat-relevant criteria presented in Chapter 5

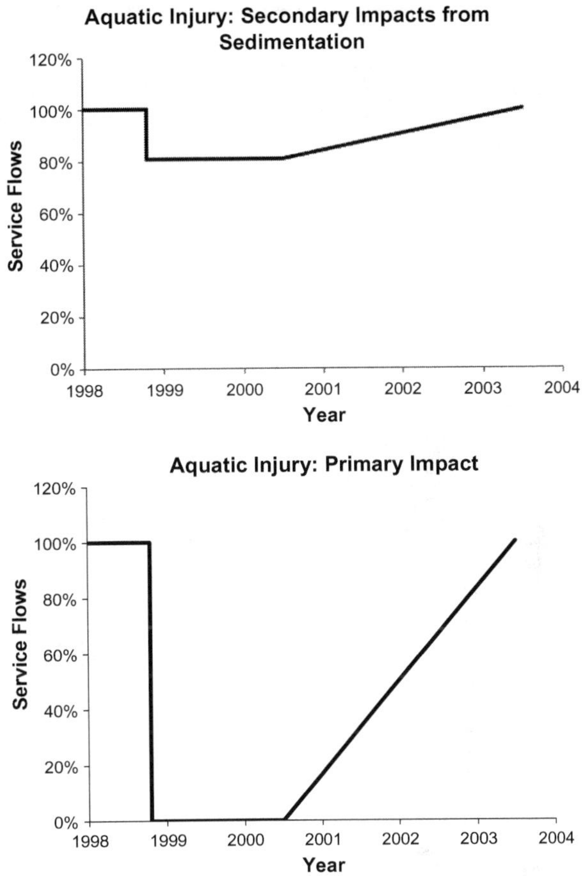

Fig. 9.3 Illustration of percent ecological service loss and recovery within primary and secondary impact zones for aquatic habitats

year of the project, the service flows from preservation in the first year are also 5%. If the marginal probability of development in the second year is also 5%, then the service flows from preservation are represented by the cumulative probability of development, or [1 − (1–5%)2] = 9.75%. The probability of development increases in successive years and eventually approaches 100%. For this case study, we adopted the assumption of a 5% annual probability of development, with the resulting increase in ecological services as illustrated in Fig. 9.4.

We performed a sensitivity analysis to demonstrate the influence of the estimated annual probability of development on the total DSHaYs of benefits that would result from preservation of 1 ha of forest habitat. The sensitivity analysis, the results of which are illustrated in Fig. 9.5, assumed that the project would be initiated in 2010 and that benefits would be realised over a 100-year period. As can be seen in the figure, project benefits are highly sensitive to the estimated annual probability of development. A 1% annual marginal probability of development corresponds to 5 DSHaYs of credit per hectare of preserved habitat; a 20% annual probability of

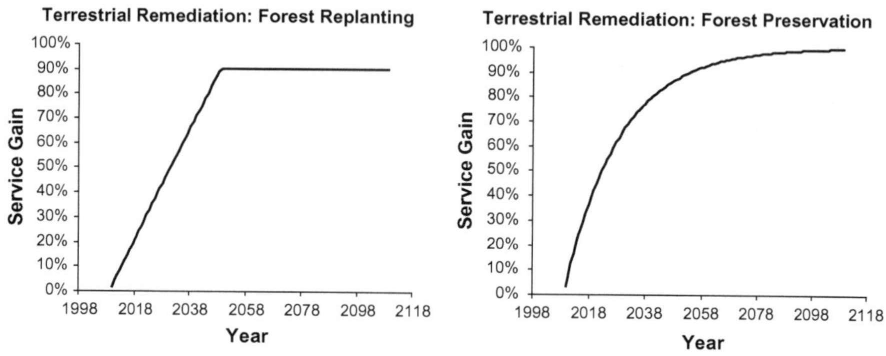

Fig. 9.4 Illustration of percent ecological service gain for terrestrial remediation, including forest replanting (left panel) and forest preservation (right panel)

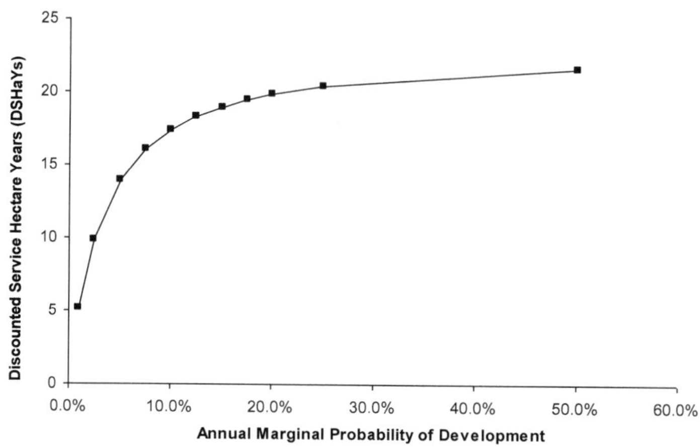

Fig. 9.5 Sensitivity analysis showing the relationship between the annual marginal probability of development and the ecological credit (in DSHaYs) generated by each hectare of forest habitat preserved from future development

development corresponds to 20 DSHaYs of credit per hectare. The selected assumption of a 5% probability of development each year leads to a level of credit falling between these two amounts. The higher the probability of development, the larger the DSHaYs becomes.

Table 9.9 shows the credits associated with the two forest remediation options. Assuming a 5% annual probability of development, forest preservation provides slightly more benefits than forest replanting.

Calculating the scale of required compensatory remediation involves dividing the number of DSHaYs of forest-normalised debit by the DSHaYs of credit generated by each hectare of remediation. This yields the total hectares of remediation

Table 9.9 Vistula River Crossing—calculated remediation credits (DSHaYs) for each hectare of forest remediation

Forest replanting alternative	
Percent service gain	90
Years to full service gain	40
Assumed year of project implementation	2010
Final year of quantified benefits	2110
Remediation credits (DSHaYs/ha)	11
DSHaYs lost (from Table 9.6)	40
Total forest replanting needed	40/11 = 3.6
Forest preservation alternative	
Annual marginal probability of development (%)	5
Assumed year of project implementation	2010
Final year of quantified benefits	2110
Remediation credits (DSHaYs/ha)	14
DSHaYs lost (from Table 9.6)	40
Total forest preservation needed	40/14 = 2.9

required for each alternative. A total of 2.9 ha of forest preservation (40 DSHaYs from Table 9.6 divided by 14 DSHaYs from Table 9.9) or 3.6 ha of forest replanting (40 DSHaYs from Table 9.6 divided by 11 DSHaYs from Table 9.9) provide benefits that compensate for the residual terrestrial damages caused by the pipeline project.

9.5.2 *Aquatic Remediation*

Similar calculations were undertaken for aquatic habitat remediation using the channel improvement alternative. To calculate credits, we again assumed that projects would be implemented in 2010 and that benefits would be quantified through to 2110. We also assumed, based on literature on channel improvement remediation and discussions with local biologists, that a 30% habitat service value increase would accrue from project implementation and that benefits would be realised over a six-year recovery period. The gain in resource services is illustrated in Fig. 9.6. Based on these assumptions, we determined that approximately 6 DSHaYs of credit would be achieved for each hectare of channel remediation (Table 9.10).

Again, total hectares of remediation are estimated by dividing the number of DSHaYs of shallow-water-normalised debit by the DSHaYs of credit generated by each hectare of channel remediation. A total of 5 ha of channel remediation (30 DSHaYs from Table 9.7 divided by 6 DSHaYs from Table 9.10) is required to provide benefits that compensate for the residual aquatic damage caused by the pipeline project.

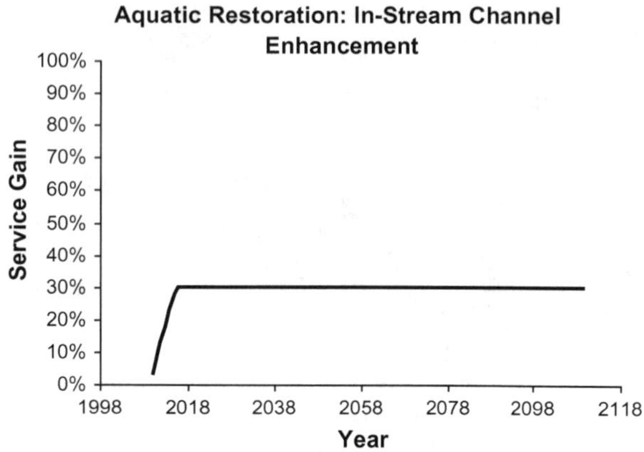

Fig. 9.6 Illustration of percent ecological service gain for aquatic remediation, consisting of in-stream channel enhancement

Table 9.10 Vistula River Crossing—calculated remediation credits (DSHaYs) for each hectare of in-stream channel enhancement

Percent service gain	30
Years to full service gain	6
Assumed year of project implementation	2010
Final year of quantified benefits	2110
Remediation credits (DSHaYs/ha)	6
Total DSHaYs loss (from Table 9.7)	30
Total in-stream channel remediation needed	30/6 = 5

9.5.3 *Calculating Total Liabilities*

To calculate total environmental liabilities associated with the Yamal Pipeline crossing of the Vistula River, the per hectare costs to implement the remediation projects must be multiplied by the total number of hectares of compensatory remediation required.

Table 9.11 provides an estimate of the costs of riparian forest remediation. These cost estimates were based on consultation with local experts in Poland. We estimated that the costs of riparian forest remediation are approximately €12,000 per hectare for the replanting options and about €7,150–9,350 per hectare for the preservation options.

Table 9.12 provides an estimate of the costs of channel improvements for aquatic habitats. These cost estimates relied on the experience of local biologists and a search of published Polish sources. The cost of channel improvement is estimated to be approximately €31,200 per hectare.

Multiplying the remediation costs shown in Tables 9.11 and 9.12 by the amount of remediation required to compensate for interim losses (Tables 9.9 and 9.10) yields total liabilities: €43,056 Euros for the forest replanting option; €20,735–27,115 for the forest preservation option and €152,880 for aquatic damage. Total liabilities depend on which terrestrial remediation option is selected. Assuming

Table 9.11 Vistula River Crossing—estimated costs of riparian forest remediation

Cost element	Forest replanting alternative (€/ha)	Forest preservation alternative (€/ha)
Planning and design	1,500	500
Permitting	1,000	–
Land acquisition	1,500	4,000–6,500
Project implementation	1,800	–
Operation and maintenance (40 years)	2,400	500
Monitoring	1,000	1,000
Subtotal	9,200	6,500–8,500
Contingency (30% for replanting; 10% for preservation)	2,760	650–850
Total	11,960	7,150–9,350

Table 9.12 Vistula River Crossing—estimated costs of aquatic channel remediation

Cost element	Cost (€/ha)
Planning and design	2,500
Permitting	1,000
Land acquisition	1,000
Project implementation	17,500
Operations and maintenance (6 years)	1,000
Monitoring	1,000
Subtotal	24,000
Contingency (30%)	7,200
Total	31,200

implementation of forest preservation alternative, which appears to be the more cost-effective option based on information available at this time, total liabilities would be approximately €170,000–180,000 (Table 9.13). Assuming implementation of the forest replanting, total liabilities are approximately €196,000.

Table 9.13 Vistula River Crossing—total estimated liabilities associated with interim losses from the pipeline construction

Habitat	Remediation alternative	Total liability (€, 2007 prices)
Terrestrial damage	Forest replanting	43,056
	Forest acquisition/preservation	20,735–27,115
Aquatic damage	In-stream channel rehabilitation	152,880
Total: option 1	Terrestrial replanting + channel improvements	195,935
Total: option 2	Terrestrial preservation + channel improvements	173,615–179,995

9.6 Uncertainties

The case study analysis presented here contains a number of uncertainties. Each key input assumption used to calculate debits and credits is somewhat uncertain. There are also uncertainties associated with the equivalency analysis. These include the service benefits of remediation alternatives and annual probabilities of development for the forest preservation option. In addition to these uncertainties, several categories of potential damage were omitted from our analysis because data were not available to support their inclusion. These omitted categories include the potential disturbance of migratory fish services during the project construction period, potential post-construction inhibition of fish migration resulting from sonic resonance of the pipeline, and potential post-construction loss of aquatic services associated with sediment destabilisation caused by pipeline resonance. Had these omitted damage categories been included in our calculations, environmental liabilities associated with the pipeline project likely would have been greater.

9.7 Conclusions

Construction of the Yamal Gas Pipeline crossing of the Vistula River caused substantial environmental damage to protected aquatic and terrestrial habitats during its two-year construction period. Terrestrial damages were caused by complete removal of habitat during excavation. The excavated area was replanted immediately following construction. The replanting was successful, thereby limiting interim loss damages. Aquatic damages were caused by excavation of the stream bottom, deposition of excavated dredge spoils into the channel, and sedimentation impacts from both the excavation and deposition processes. No primary remediation actions were performed for the aquatic environment, although biologists estimated that the river recovered within approximately three years of construction.

We performed habitat equivalency analyses to illustrate how interim losses could be calculated on a habitat basis, even for projects with relatively short-lived direct impacts. We calculated environmental damage debits within both the primary impacts zone caused by excavation and a secondary impacts zone associated with terrestrial and aquatic disturbance. The use of habitat scalars enabled us to normalise damages to riparian forest habitat for the terrestrial impacts and shallow-water habitat for aquatic impacts.

Compensatory remediation projects were selected to provide benefits to riparian forest and shallow-water habitat types. For riparian forest remediation, two options were deemed feasible: forest replanting and forest preservation. The preservation option generated slightly more benefits than the replanting option, but we found this outcome to be highly sensitive to the estimated annual probability of development the remediation option would prevent. After applying a series of project evaluation criteria, in-stream channel improvements were determined to be the preferred remediation option for aquatic remediation. Total cost of compensatory remediation was estimated to be approximately €170,000–200,000.

Although construction of the Yamal Gas Pipeline crossing occurred prior to passage of the ELD and therefore would not be subject to ELD, this case study demonstrated how environmental damages associated with similar projects could be calculated.

Acknowledgements The study team thanks Sandy Luk, Professor Andrzej Przystalski, Dr. Wiesław Cyzman, Dr. Krzysztof Kasprzyk, and Dr. Eng. Wiesław Wiśniewolski for expertise, advice, and support.

Chapter 10
Ex-Ante Analysis of a Hypothetical International Road Construction Project in Poland

Joshua Lipton, Zenon Tederko and Eric English

Abstract This case study uses equivalency analysis to identify and quantify environmental damage, remediation benefits and compensatory liabilities for the construction of a hypothetical international road in northeastern Poland. Habitat equivalency analysis (HEA) was used as an evaluative approach to comparing the potential environmental damages associated with two alternatives. Application of HEA on an *ex-ante* basis enabled us to compare the cost effectiveness of the two alternatives considering their relative future environmental damages. Under a simple base case, environmental damages for the proposed Route G alternative were somewhat greater than for the Route N alternative. When we considered potential wide-scale ecosystem damages using a probabilistic approach, however, environmental damages for Route G were considerably greater than for Route N. When this probabilistic approach was expanded further to consider the relative scarcity of the extremely rare alkaline fen habitat that could be lost, the cost of necessary remediation increased considerably to over €11 billion. This case study is an illustration of how equivalency analysis can be used to implement the Habitats, Environmental Impact Assessment, and Strategic Environmental Assessment Directives.

Keywords Habitat equivalency analysis · Alkaline fen habitat
Habitats directive · Infrastructure construction · Poland

J. Lipton (✉)
Abt Associates, 1881 Ninth Street, Suite 201, Boulder, CO 80302, USA
e-mail: Josh_Lipton@abtassoc.com

Z. Tederko
Pro-Biodiversity Business Services Ltd, Al. Komisji Edukacji Narodowej 46/1, 02-797 Warsaw, Poland
e-mail: zenon.tederko@gmail.com

E. English
Bear Peak Economics, 623 Maxwell Avenue, Boulder, CO 80304, USA
e-mail: Eric.English@bearpeakeconomics.com

J. Lipton et al. (eds.), *Equivalency Methods for Environmental Liability*,
https://doi.org/10.1007/978-90-481-9812-2_10

10.1 Introduction

This case study assesses the potential environmental damages associated with the construction of a hypothetical international road project. We assume that the hypothetical road project would occur in northeastern Poland. The case study provides an illustration of equivalency analysis in an *ex ante* context (analysis before the damage occures).

For the Environmental Impact Assessment Directive (EIAD) and the Strategic Environmental Assessment Directive (SEAD) to apply, the road building project needs to be a public or private project that is likely to have significant effects on the environment[1] or a plan or program that is subject to preparation and/or adoption by an authority at national, regional, or local level or that is prepared by an authority for adoption, through a legislative procedure by Parliament or Government and is required by legislative, regulatory, or administrative provisions.[2] One or both of these characteristics are likely to apply to major road building projects. Therefore, the road building project in this case study is assumed to fall under the provisions of these Directives. It is also assumed that mitigation measures will be required. However, it is also assumed that the project will consider different routing possibilities to reduce the damage[3] (BBOP and UNEP 2010).

Similarly, an appropriate assessment under Article 6(3) of the Habitats Directive (HD) is required in relation to plans or projects likely to have a significant effect on Natura 2000 sites. If there are likely to be significant adverse effects, development is only allowed for imperative reasons of overriding public interest.[4] In such a case, compensatory measures are required. However, in relation to priority natural habitat types and/or priority species, imperative reasons of overriding public interest can only be those relating to human health or public safety, beneficial consequences of primary importance for the environment or, further to an opinion from the European Commission, to other imperative reasons of overriding public interest.[5]

For the purpose of this case study, it is assumed that the necessary reasons of overriding public interest exist and the European Commission's opinion has confirmed this in the case of priority habitats or species. Consequently, the case study assumes that the construction falls within the overall scope of applicable European Union Directives and that equivalency analyses may be applied on an *ex ante* basis.

Section 10.2 describes the project. Section 10.3 presents the initial evaluation of the impacts of the incident. Section 10.4 quantifies debits from environmental damage, while credits from remediation are quantified in Sect. 10.5. Scaling of

---

[1]Article 1, EIAD.

[2]Article 2(a), SEAD.

[3]Mitigation hierarchy—used for Environmental Impact Assessments and biodiversity offsets—requires that projects first avoid, then minimise impacts, and restore and finally offset residual impacts.

[4]Article 6(4), HD.

[5]Ibid.

remediation to sufficiently offset damages is discussed in Sect. 10.6. The chapter ends with a brief discussion of conclusions (Sect. 10.7).

10.2 Description of the Incident

This case study involves the anticipated environmental damage associated with construction of a hypothetical international highway. Specifically, we address a situation in which a highway bypass is planned that would link two cities: City A and City B. Two alternative routes are proposed. Route G would involve construction of the road through pristine wetlands in a river valley. Two Natura 2000 sites (a Primeval Forest Special Protection Area (SPA) and potential Primeval Forest Special Area of Conservation (SAC)) lie within the path of Route G. Route G would include construction of an elevated causeway some 500 m in length above the wetlands and the river valley. The alternative route, N, would bypass the Natura 2000 sites and would be constructed within a pre-existing corridor containing electricity transmission lines.

Both routings would likely adversely affect the river valley. Route G, however, would affect important wetlands and Natura 2000 sites. Table 10.1 summarises key characteristics of the alternative construction routes.

10.3 Initial Evaluation of Affected Habitats and Species

10.3.1 Protected Areas

The hypothetical study are contains several legally protected habitats. These include a Primeval Forest SPA site, covering nearly 120,000 ha; a HD SAC site, covering more than 120,000 ha and proposed as a Site of Community Importance; and a Primeval Forest/Important Bird Area of European Union Importance, covering nearly 135,000 ha.

The Primeval Forest is an extensive complex of relatively dense, old-growth forest. It lies on a postglacial, sandy plateau that is 100–140 m above sea level, with relatively numerous preserved glacial basins and postglacial channels, which are

Table 10.1 International road construction—characteristic of Routes N and G

	Route N	Route G
Total road length (km)	41.25	41.35
Length of road within Natura 2000 sites (km)	1	12
Length of river valley at crossing (m)	130	500
Length of elevated bridge/causeway at river crossing (m)	150	517.34
Construction costs (million €)	260	240

filled by lakes and bogs. Water bodies totalling 5% of the site surface are especially numerous in the western, central, and northern part of the SPA, with a large lake complex east of City B.

The majority of the site (~ 85%) is forested, with a few open patches used for agriculture, mostly as grasslands. Coniferous forests, predominantly pine stands, prevail. There are also patches of well-preserved wet and swampy coniferous forests growing on bogs. However, the lowest, very wet, places in the river valleys and around the lakes are typified by riparian wetlands. In some places, dry forests and subboreal swampy birch forests are extant. The Primeval Forest is situated within the range of the continental climate, and many boreal and subboreal species are present.

10.3.2 Vegetation and Habitats

An extensive network of wetland habitats within the river valley is included within the Natura 2000 network as part of the SPA Primeval Forest under the Wild Birds Directive (WBD). These habitats also are included on the Shadow List of sites of Community Importance under the HD.

Twenty-four habitat types of community importance (including eight priority habitats) listed in Annex I of the HD cover approximately 17% (204 km^2) of the Primeval Forest. Extensive patches of bog woodland (Eurpoean Union Habitat Code 91D0*), mixed pine-birch stands, and other bog and fen habitats (codes 7110, 7140, 7150, 7210, 7230) highlight the importance of the site for listed habitats.

Most of the peatland area is covered by vegetation types included in Annex I of the HD. Approximately 100 ha (more than 15% of the valley) is occupied by alkaline fen (code 7230) with sedge-moss rich fen vegetation. Nearly 300 ha are covered by bog woodland (code *91D0), a priority vegetation type.

The extensive open sedge-moss fen communities are the most valuable habitat of the river valley. They have permanently high water levels and are largely free of invading willow or birch shrubs. The plant communities are dominated by sedges (*Carex rostrata*, *C. diandra*, *C. limosa*, *C. lasiocarpa*, *C. chordorrhiza*) and brown mosses (including *Drepanocladus* s.l., *Tomenthypnum nitens*, *Calliergon giganteum*, *Calliergonella cuspidata*, *Aulacomnium palustre,* and locally calcitolerant *Sphagnum* spp.). These communities belong to the mesotrophic small sedge-brown moss vegetation, with a high number of calcicole species from the *Caricion davallianae* assemblage.

The vegetation zones in the valley include aquatic plant communities close to the riverbed, reed beds, tall sedge communities, an extensive zone of sedge-moss fen communities of the *Scheuchzerio-Caricetea* class, bog woodland and pine-birch shrubs classified as the *Thelypteridi-Betuletum pubescentis* association, and spruce forests or peat or alder swamp forests close to the mineral slopes of the valley.

The sedge-moss communities, covering more than 100 ha, are the most valuable habitat of the river valley. They have permanent high water levels and, in most parts, are free of encroaching willow or birch shrubs. They are dominated by small sedge communities with *Carex rostrata, C. diandra, C. limosa, C. lasiocarpa, C. chordorrhiza*, and 'brown mosses,' mostly *Drepanocladus* s.l., *Tomenthypnum nitens, Calliergon giganteum, Calliergonella cuspidata*, and *Aulacomnium palustre* with an addition of Sphagna at places.

10.3.3 Threatened Species

Several rare and threatened species occur within the study area. The rarest species of vascular plants occurring in the valley are: musk orchid (*Herminium monorchis*), three species protected by the HD (marsh saxifrage (*Saxifraga hirculus*), fen orchid (*Liparis loeselii*), lady's slipper orchid (*Cypripedium calceolus*)), dwarf birch (*Betula humilis*), Jacob's ladder (*Polemonium caeruleum*), adder's-mouth orchid (*Malaxis monophyllos*), slender cotton-grass (*Eriophorum gracile*), and cotton deergrass (*Baeothryon alpinum*). Among bryophytes, many are relict species, for example, *Paludella squarrosa*.

At the time of conducting this case study, the open fens of the river valley were a last resort for many plant species that are endangered in Poland and the rest of Europe. As many as 14 vascular plant species were included in *Poland's Red Data Book of Plants* (e.g., *Eriophorum gracile, Carex chordorrhiza, Baeothryon alpinum, Herminium monorchis*); 32 species of vascular plants, mosses, and liverworts were listed in the Polish 'red list' (e.g., *Meesia triquetra, Paludella squarrosa, Tomentypnum nitens*); and 75 species were under protection in the country. The valley was the only site in Poland where musk orchid *(Herminium monorchis)* occurs. It accommodated the most numerous and best-preserved Polish populations of two HD species: *Liparis loeselii* and *Saxifraga hirculus.* Considerably fewer protected plants occur in proximity to the planned Route N.

In the Primeval Forest SPA, at least 42 breeding bird species listed in Annex I of the WBD were found. In addition, 12 species found in the SPA were included in the list of threatened birds in *Poland's Red Data Book of Animals* (2001). For eight Annex I WBD species, the Primeval Forest was one of the 10 most important breeding sites in Poland, supporting >1% of their national populations. These species (referred to as key species) included black stork, honey buzzard, lesser spotted eagle, capercaillie, grey-headed woodpecker, white-backed woodpecker, three-toed woodpecker, and red-breasted flycatcher. Further, the site supported a large breeding population of crane (just below 1% of national total). In addition, the Primeval Forest provided breeding habitat for a number of rare raptor species including the short-toed eagle, black kite, red kite, and white-tailed eagle. As with vegetation, considerably more protected avian species occurred within proximity to Route G than Route N.

The Primeval Forest habitat is also home to the following five species of mammals listed in Annex II of the HD: wolf, lynx, otter, beaver, and pond bat. In addition, the local population of elk is of considerable importance. Together with other neighboring natural forests, the Primeval Forest represents the largest continuous forest tract in northeast Poland and is of key importance in maintaining the largest viable metapopulations of lynx and wolf in the lowlands of Poland and Central Europe.

10.3.4 Landscape Values

In addition to being a unique mire ecosystem, the river valley offers landscape values. These landscape values derive from the unspoiled nature of the habitat associations, the low degree of human impacts within the overall watershed, the extensive forest buffer surrounding the wetland habitats, and the large number of rare and critical plant species that occur across the various habitat gradients.

10.3.5 Preliminary Identification of Potential Damages: Stressors

Construction of the road would involve a number of stressors that could damage species, habitats, and landscape values. Environmental damage could occur during both the construction and operation phases of the project. Examples of anticipated stressors during the construction phase include:

- Direct loss of habitat from construction;
- Disturbance of fauna due to human presence, equipment operations, noise, and light;
- Creation of barriers for animal movements—both for migration and normal dispersal movements;
- Temporary or permanent changes in hydrology (both groundwater and surface water);
- Vibrations from driving piles;
- Shading from platforms and bridges;
- Primary and secondary dust from excavation, traffic, and equipment;
- Air pollution;
- Sedimentation; and
- Altered patterns of water runoff and sediment yields in local drainage basins.

In addition to the above stressors, examples of anticipated stressors during road operations include:

- Disturbance from invasive roadside species;
- Facilitated spread of pathogens and diseases, as well as of exotic and pest species along roadways;

- Increased human access affecting wildlife, fire, and other disturbance to sensitive habitats and species;
- Vibrations caused by heavy traffic;
- Introduction of traffic or/and increase of traffic intensity and related pollution and risks;
- Salt and other ice-control chemicals;
- Pollution in road runoff;
- Noise and light; and
- Accidents with hazardous products.

10.3.6 Potential Impacts Associated with the Road

To evaluate potential adverse effects from the road project, we first considered the sensitivity and fragility of the unique river mire ecosystem. Specifically, it is important to understand that integrity of these peatland ecosystems relies on the interconnectivity of hydrological and vegetation processes. The composition of vegetation communities determines the type of peat that will be formed and the nature of its hydraulic properties. Site hydrology, in turn, determines which plants will grow, whether peat will be created, and decompositional processes. The peat structure and the physical relief determine how the water will flow and fluctuate. These close interrelationships imply that when any one of these components changes, the others will be affected as well.

The river mire in the study area is categorised as a sloping mire, where the water level forms an inclined plane and water flow is mainly horizontal. The laterally flowing water is retarded by vegetation and peat. Vegetation growth and peat accumulation actively cause a rise of the water table in the mire and often also in the catchment area.

The specific type of sloping mire found within the study area is known as a percolation mire. Percolation mires are found in areas where there is adequate water supply evenly distributed over the year. As a result, the water level in the mire is almost constant; dead plant material reaches the permanently waterlogged zone quickly and is subject to aerobic decay only for a short time. Consequently the peat remains weakly decomposed and highly permeable so that the water flows through a considerable part of the peat body. Because the only weakly decomposed peat also remains elastic, the mire surface can oscillate with changing water supply, leading to very constant water levels relative to the surface and very stable conditions for peat formation. With growing peat thickness, this mire oscillation capacity increases, and the mire becomes less susceptible to absolute water level fluctuations.

Percolation mires are normally fed by groundwater because in most climates only large catchment areas can guarantee the necessary large and continuous water supply. Groundwater-fed percolation mires contain a high diversity of strongly specialised species that are not found in other habitats. These often rare and threatened peatland plants have special adaptations to deal with the extreme lack of phosphate that is made unavailable by groundwater-derived iron and calcium.

Along the river, a narrow zone is regularly flooded and, consequently, is more nutrient rich. Furthermore, rainwater 'lenses' that are extremely poor both in minerals and nutrients have developed locally in the center of the valley. The smooth and fine-scaled gradients between these three conditions (groundwater-, floodwater-, and rainwater-fed) lead to numerous intermediate situations and a high diversity of ecological niches.

Within the temperate zone of Europe, the study site is the best remaining example of a percolation mire. The river valley, furthermore, has excellent prospects for long-term conservation. The hydrogeological system bears no signs of anthropogenic disturbances. Human impact has been low not only in the valley but also in its catchment, which is largely forested and in low-intensity use. The surrounding forest forms a buffer zone that limits the influence of the nearby agricultural land.

The good hydraulic, hydrologic, and hydrogeochemical conditions of the river mire guarantee slow successional processes and stable habitat conditions that offer a unique chance to preserve endangered and protected habitats and species without active (and expensive) management.

However, fens are among the most sensitive ecosystems in Europe, susceptible to degradation through any interference of their local and regional hydrological regime. There has already been large-scale and severe degradation of fen systems in Poland and throughout Europe. Given the fragile hydrological equilibrium, road development could create irreversible damage to the fens of the river valley. Although mitigation measures may be effective in reducing the impact, they would not ensure preservation of the key functional aspects of this system.

10.3.7 Potential Impacts to Fauna

Road development may adversely affect resident and migratory fauna through a number of processes, including:

- Increased mortality from collisions with vehicles;
- Decreased densities in areas adjacent to the road as a result of increased disturbance (noise, visual and human disturbance, pollution), which reduce habitat quality;
- Loss of supporting habitat;
- Altered animal behaviour through changes in activity patterns, spatial behaviour, and increased stress; and
- Increased fragmentation of animal populations caused by the barrier effects of roads. Fragmentation can cause increased probability of extinction of local or isolated populations. Roads with heavy traffic can seriously impair animal movements (dispersal and migrations) and gene flow across both sides of the expressway.

10.3.8 Potential Habitat Impacts

Habitat impacts associated with road development may include:

- Considerable and far-reaching changes in freshwater habitats caused by substantially altered patterns of water runoff and sediment yields in local drainage basins, coupled with increased chemical pollution and light pollution;
- Direct habitat loss from construction; and
- Decreased habitat quality from construction and operation.

10.3.9 Potential Impacts: Landscape Fragmentation, Ecological Integrity, and Ecological Connectivity

The habitats affected are likely to directly serve as the dispersal corridor for large mammals such as wolf, lynx and elk. In a broader spatial context, this dispersal corridor is the main tract supporting the connectivity of continuous forest tracts of the South Baltic basin (Lithuania, Latvia, Belarus, Russia) and fragmented forests of Central Europe.

One of the main requirements for proper functioning of wolf and lynx populations is the maintenance of long-distance dispersal movements of individuals. If Route G is selected, the substantial increase in traffic volume on road sections placed in dispersal corridors is likely to seriously impair dispersal. Animals not only avoid busy roads but also suffer heavy mortality while crossing them. Limiting the dispersal possibilities would reduce the probability of animals colonising in new areas. Even more importantly, it could also cause the isolation of existing local populations that would then be exposed to higher risk of decline and extinction.

The prevention of dispersal may be a particularly important threat to the lynx population of northeastern Poland, which is almost entirely isolated from the bulk of the Baltic basin population. Road mortality is invariably identified as one of the most important sources of lynx mortality across Europe. Additionally, lynx numbers in Poland have been declining since 1989. Therefore, the lynx population of northeast Poland is particularly vulnerable and requires efficient conservation measures, including protection of dispersal routes.

10.3.10 Anticipated Temporal Extent of Damage

The construction period for both alternatives is estimated to be 24 months. Consequently, direct effects associated with construction (noise, dust, equipment) are anticipated to last for two years, with recovery assumed to occur within 1 year. Habitat losses caused by the road, once constructed, are anticipated to continue in perpetuity.

10.3.11 Primary Remediation Undertaken

For the purposes of this case study, we assume that no primary remediation will be undertaken as part of the project. Rather, we use equivalency analysis to determine the scale of the offsetting mitigation that would be required for the two road alternatives.

10.3.12 Preliminary Identification of Potentially Affected Services

The road construction project is likely to result in a range of ecological losses. Table 10.2 identifies potentially affected habitats, resources, and services within both the direct road path (approximately 60 m wide), which we refer to as the primary impact zone, and within 1 km on either side of the road, which is the secondary impact zone.

Table 10.2 International road construction—potentially affected habitats, resources, and services

Primary impact zone (60-m road width)	Secondary impact zone (2-km-wide buffer)
Permanent loss of protected plants, including loss of individuals and loss of supporting habitat	Temporary loss of avian breeding habitats
Permanent loss of avian breeding habitats	Temporary loss of avian feeding sites
Permanent loss of avian feeding sites	Temporary loss of mammalian habitat
Permanent loss of mammalian habitats	Temporary loss of amphibian habitat
Permanent loss of amphibian habitats	Temporary loss of reptile habitat
Permanent loss of reptile habitats	Temporary loss of fish habitat
Permanent loss of fish habitats	Temporary loss of insect habitat
Permanent loss of insect habitats	Temporary loss of habitat integrity and connectivity
Permanent fragmentation of habitats and landscapes	Temporary loss of greenhouse gas sink (sequestration) ability
Permanent change of hydrological regime along the road	Temporary loss of recreation usage
Loss of hydrological stability of fens	Temporary separation from arable fields
Decrease in the hydraulic conductivity of the substrate	
Permanent fragmentation of agricultural land	

10.3.13 Potential Social, Economic, and Transboundary Issues

For the purposes of this case study, we focus on the loss of ecological resources, specifically habitat services. Losses of social, economic, or recreation-related services are assumed to be modest. The case study area is sparsely populated, and the road project would not substantially affect social or economic values. No significant social, economic, or transboundary issues are anticipated for either road alternative.

10.4 Quantifying Debits from Environmental Damages

We used HEA to compare potential environmental damages from the two alternative routes including both primary and secondary impact zones as described above (see Table 10.2). Although the project would potentially affect a number of different habitat types, we simplified our HEA analysis by pooling similar types into five discrete habitat assemblages:

- For Forest Habitats, we evaluated three habitat groups: bog forests (inclusive of the bog woodland assemblage, habitat code 91D0), alluvial forests (inclusive of alluvial forests with *Alnus glutinosa* and *Fraxinius excelsior*, habitat code 91E0), and oak forests (inclusive of Galio-Carpinetum oak-hornbeam forests, habitat code 9170).
- For Wetland Habitats, we pooled individual habitat types into two general assemblages: alkaline fens (inclusive of alkaline fens and transition mires, codes 7230 and 7140) and common wetlands (e.g., habitat code 3260).

As discussed above, road construction could adversely affect a number of different ecological services. Consequently, a variety of ecological metrics could be used to describe debits (and credits). However, rather than apply individual metrics to different potential categories of loss, we employed an overall habitat-integrity metric to describe changes across a range of ecological services. This metric was based on the professional judgment of local scientists and experts. It describes an overall *gestalt* view of habitat integrity and considers impacts to local flora and fauna, potential hydrological impacts to wetland habitats, and habitat fragmentation/connectivity. Although semiqualitative multi-resource metrics that rely on professional judgment may not be ideal, particularly in *ex post* analyses, use of such metrics may be warranted in certain *ex ante* analyses for which quantitative data on environmental damage may not be available, particularly in situations in which alternatives are being considered and contrasted.

Finally, we employ habitat scalars in our HEA to provide for equivalency scaling between habitat assemblages. We also considered uncertainties in potential future outcomes in this *ex ante* case by performing a probabilistic analysis of alternative scenarios.

Habitat service losses were quantified for Routes N and G. For each route, ecological service losses were defined for all habitat assemblages for both primary and secondary impact zones. Tables 10.3 and 10.4 present the quantification of affected habitats within the primary and secondary impact zones and related service loss assumptions for Routes G and N, respectively.

As shown in the above tables, the number of hectares that would be impacted is significantly higher for Route G than for Route N. The greatest difference with respect to forest habitats occurs in bog forests. Specifically, the secondary impact zone for Route G includes 79 ha of bog forest, while secondary impact zone for Route N includes only 0.4 ha of bog forest. The greatest difference with respect to wetland habitat occurs in alkaline fens. Secondary impacts for Route G include 35 ha of alkaline fens, while no alkaline fen habitat is impacted for Route N.

Table 10.3 International road construction Route G—ecological service loss assumptions by habitat type and impact zone

Habitat type	Damaged area (ha)		Ecological service loss (%)		Duration of loss during construction (years)	Recovery period (years)	
	Primary	Secondary	Primary	Secondary		Primary	Secondary
Forest							
Bog	1	79	100	50	2	None	5
Alluvial	0	15	na	50	2	None	5
Oak	5	167	100	50	2	None	5
Wetland							
Alkaline fens	1	35	100	50	2	None	5
Common wetland	1	7	100	50	2	None	5

Table 10.4 International road construction Route N—ecological service loss assumptions by habitat type and impact zone

Habitat type	Damaged area (ha)		Ecological service loss (%)		Duration of loss during construction (years)	Recovery period (years)	
	Primary	Secondary	Primary	Secondary		Primary	Secondary
Forest							
Bog	0	0.4	na	50	2	na	5
Alluvial	0	5	na	50	2	na	5
Oak	4	139	100	50	2	None	5
Wetland							
Common wetland	0.1	8	100	50	2	None	5

Primary loss continues in perpetuity for all habitat types where a primary loss occurs
Secondary loss continues for seven years for all habitat types

Although the extent of habitats to be impacted differs for the two routes, the pattern of primary and secondary impact zone service losses are consistent for both alternatives. Service losses in the primary impact zone are assumed to begin with a loss of 100% at the time of construction and continue in perpetuity. In the secondary impact zone, initial losses are assumed to be 50%, which reflects an average of losses that vary with distance from the road. Specifically, it is assumed that losses are 100% immediately adjacent to the road and then decline linearly to zero at the edge of the secondary impact zone. This is equivalent to an average loss of 50% extending 1 km in either direction from the road. The initial 50% average loss is assumed to continue during the two-year construction period. Once construction ends, losses in the secondary impact zone decline to zero over a period of five years. Loss assumptions for the primary and secondary impact zones are illustrated graphically in Fig. 10.1.

Using the information in Tables 10.3 and 10.4, we calculated total debits associated with habitat damages for the two routes. We used 2008 as the base year, combined with a 3% discount rate, to calculate the present value of ecological service losses. Present value service losses were expressed in Discounted Service Hectare Years (DSHaYs). Table 10.5 presents an example of the debit calculations for the damages to the oak forest habitat in the primary impact zone for Route N. To calculate present value, the constant service loss of 4.0 ha in each year is multiplied by the discount factor in each year, which declines through time. The result is shown in the final column of Table 10.5. The sum from 2008 to perpetuity is 134 DSHaYs.

The terminal value represents combined discounted losses into perpetuity, starting in the year 2108. It is calculated by dividing annual losses by the annual discount rate, then multiplying by the discount factor for 2108 ((3.9/ 0.03) × 0.05 = 6.6).

Similar calculations were undertaken for other habitat types. Tables 10.6 and 10.7 summarise the results of the total debit calculations for Routes G and N, respectively. Although the area affected in the primary impact zone is considerably smaller than the area affected in the secondary impact zone for each habitat and route, the difference in

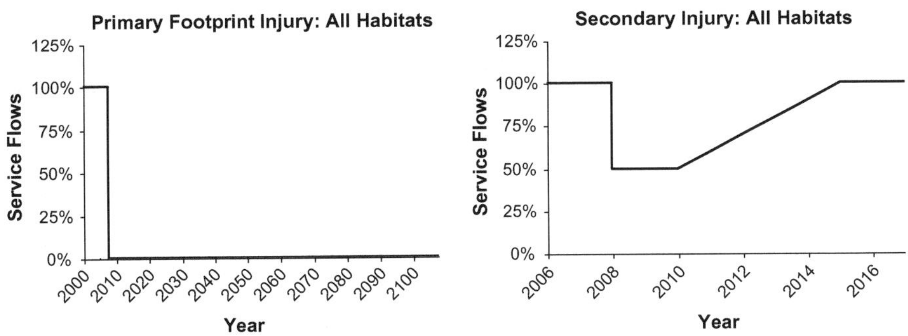

Fig. 10.1 Illustration of ecological service flows for primary and secondary impact zones of all habitats for Route G and Route N

Table 10.5 International road construction Route N—HEA calculations for 4 ha of oak forest in primary impact zone

Year	Annual ecological service loss (%)	Total annual service loss (ha)	Discount factor	Total annual discounted service loss (DSHaYs)
2008	100.0	3.9	1	4.0
2009	100.0	3.9	0.97	3.8
2010	100.0	3.9	0.94	3.7
2011	100.0	3.9	0.92	3.6
2012	100.0	3.9	0.89	3.5
2013	100.0	3.9	0.86	3.4
2014	100.0	3.9	0.84	3.3
2015	100.0	3.9	0.81	3.2
2016	100.0	3.9	0.79	3.1
⋮	⋮	⋮	⋮	⋮
2107	100.0	3.9	0.05	0.2
2108	100.0	3.9	0.05	0.2
2108+	100.0	3.9	0.05	6.6[a]
Total				134

Notes [a] The terminal value represents combined discounted losses into perpetuity, starting in the year 2108. It is calculated by dividing annual losses by the annual discount rate, then multiplying by the discount factor for 2108 ((3.9 / 0.03) × 0.05 = 6.6).
To shorten the table, some of the results were omitted and substituted by the ellipsis

Table 10.6 International road construction Route G—total DSHaYs of debit

Habitat	Impact zone		
	Primary	Secondary	Total
Forest			
Bog	28	150	178
Alluvial	0	28	28
Oak	188	317	505
Wetland			
Alkaline fens	26	66	92
Common wetland	18	13	31

Table 10.7 International road construction Route N—total DSHaYs of debit

Habitat	Impact zone		
	Primary	Secondary	Total
Forest			
Bog	0	1	1
Alluvial	0	10	10
Oak	134	264	398
Wetland			
Alkaline fens	0	0	0
Common wetland	3	15	18

calculated service losses between the two zones is less significant. This is because losses in the primary impact zone continue in perpetuity, which causes significant total losses through time even for relatively small areas of impact.

In order to combine losses across habitats, the relative value of different habitats was addressed using habitat scalars. Because remediation projects are available for bog forest and common wetlands, losses were converted to these habitat types. Specifically, all forest habitat losses were converted to bog forest losses, and all wetland losses were converted to common wetland losses. These calculations are shown in Tables 10.8 and 10.9.

As shown in Table 10.8, the habitat scalars used for bog forest, alluvial forest, and oak forest were 1.0, 0.5, and 0.33, respectively. This indicates that 2 ha of alluvial forest are equivalent to 1 ha of bog forest and that 3 ha of oak forest are equivalent to 1 ha of bog forest. As shown in Table 10.9, the habitat scalars for alkaline fens and common wetland are 15 and 1, respectively. This implies that 1 ha of alkaline fens is equivalent to 15 ha of common wetland. The habitat scalars were based on professional judgment related to the ecological functions of the different habitats, as well as the relative scarcity of the habitats.

Table 10.8 International road construction Route G—debits normalised to bog forest and common wetland

Habitat	Unadjusted DSHaYs	Habitat scalar	Normalised debit (DSHaYs)
Bog	178	1	178
Alluvial	28	0.5	14
Oak	505	0.33	168
Total bog forest equivalent			**360**
Alkaline fens	92	15	1,387
Common wetland	31	1	31
Total common wetland equivalent			**1,418**

Total equivalents are indicated in bold text

Table 10.9 International road construction Route N—debits normalised to bog forest and common wetlands

Habitat	Unadjusted DSHaYs	Habitat scalar	Normalised Debit (DSHaYs)
Bog	1	1	1
Alluvial	10	0.5	5
Oak	398	0.33	133
Total bog forest equivalent			**138**
Alkaline fens	0	15	0
Common wetland	18	1	18
Total common wetland equivalent			**18**

Total equivalents are indicated in bold text

To calculate debits normalised to bog forest and common wetland, we multiplied unadjusted DSHaYs for each habitat (as calculated in Tables 10.6 and 10.7) by the appropriate habitat scalar. The resulting normalised debits are shown in the last columns of Tables 10.8 and 10.9. The total forest loss for Route G is 360 DSHaYs when converted to normalised bog forest debits. The total adjusted forest loss for Route N is 138 bog forest DSHaYs. The total wetland loss for Route G is 1418 DSHaYs when normalised to common wetland debit. This was nearly 100 times greater than the 18 DSHaYs of normalised wetland loss for Route N.

10.5 Quantifying Credits from Remediation

Debits from damages to forest and wetland habitats are offset using remediation projects that restore bog forest and common wetland habitat types. Habitat credits for these remediation projects were calculated using the credit assumptions presented in Table 10.10. It was assumed that construction of an offsetting bog forest remediation project would be completed in 2010 and that it will take 100 years for the enhanced habitat to reach full maturity. Ecological services were assumed to increase linearly during that time, resulting in a final increase in ecological services of 75% in the year 2110. The present value of the increase in services is calculated using the same discounting methods applied to the habitat service debits. Specifically, a 3% discount rate is applied on an annual basis and annual discounted credits are summed over the 100-year life of the project. As shown in Table 10.10, total remediation credits for the bog forest project projected to be 6.7 DSHaYs per hectare of remediated habitat.

A similar set of assumptions was applied to determine credits for the remediation of common wetland, also shown in Table 10.10. The remediated wetland is

Table 10.10 International road construction—remediation credits for each hectare of remediated habitat

Bog forest	
Assumed year of project completion	2010
Years to full service gain	100
Percent ecological service gain	75%
Final year of quantified benefits	2110
Remediation credits	6.7 DSHaYs per ha
Common wetland	
Assumed year of project completion	2010
Years to full service gain	20
Percent service gain	75%
Final year of quantified benefits	2110
Remediation credits	17.4 DSHaYs per ha

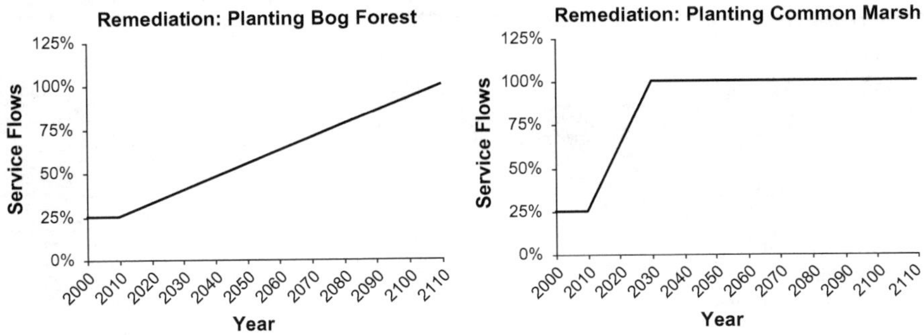

Fig. 10.2 Ecological service flows derived from forest bog and common marsh remediation projects

assumed to reach a full service increase of 75% in 20 years. Total credits for the wetland project are 17.4 DSHaYs per hectare. The service flows from the forest bog and common marsh remediation projects are illustrated in Fig. 10.2.

Remediation habitats are assumed to be constituted in year 1, and hence costs are not discounted. The numbers in the final column are based on the remediation scale that is not rounded up or down.

10.6 Scaling Complementary and Compensatory Remediation

Because outcomes are uncertain in this *ex ante* application of HEA, we present two alternative scaling scenarios. The first scenario, which we refer to as the 'base case', uses the information on debits and credits presented above to calculate the amount of remediation required to compensate for environmental damage. In the second scenario, we apply a probabilistic approach to evaluate potential damages associated with losses to highly scarce alkaline fen habitats.

10.6.1 Base Case

Our base case uses the results outlined above to calculate the amount of remediation required to offset expected losses. For Route G, the debit of 360 DSHaYs of bog forest habitat (Table 10.8) must be offset by per-unit remediation credits of 6.7 DSHaYs per hectare (Table 10.11). The required scale of the bog forest remediation project therefore is 54 ha (360/6.7 = 54). The debit of 1,418 DSHaYs of wetland habitat (Table 10.8) must be offset by the per unit remediation credits of 17.4 DSHaYs per hectare (Table 10.11). The required scale of the common wetland remediation project therefore is 81 ha (1,418/17.4 = 81). We assume, for

Table 10.11 International road construction—total remediation costs by habitat type

Habitat type	Debits (from Table 10.10) (DSHaYs)	Credits per Unit (from Table 10.10) (DSHaYs/ha)	Scale of required remediation (ha)	Remediation costs per unit (€/ha)	Total remediation costs (€)
Route G					
Bog forest	3606.7	6.7	54	10,000	537,313
Common wetland	1,418	17.4	81	10,000	814,943
Total Route G					**1,352,256**
Route N					
Bog forest	138	6.7	21	10,000	205,970
Common wetland	18	17.4	1	10,000	10,345
Total Route N					**216,315**

Total equivalents are indicated in bold text

illustrative purposes only,[6] a remediation cost of €10,000 per hectare for both habitat types. This results in a cost of €537,313 to offset forest damages and €814,943 to offset wetland damages. Thus, the total cost of remediation for Route G, under the base case analysis, would be €1,352,256.

For Route N, the debit of 138 DSHaYs of bog forest habitat must be offset by the per unit remediation credits of 6.7 DSHaYs per hectare. The required scale of the bog forest remediation project therefore is 21 ha (138/6.7). The debit of 18 DSHaYs of wetland habitat must be offset by the per unit remediation credits of 17.4 DSHaYs per hectare. The required scale of the common wetland remediation project therefore is 1 ha. Again, assuming an illustrative remediation cost of €10,000 per hectare, this results in a cost of €205,970 to offset forest damages and €10,000 to offset wetland damages. Thus, the total cost of remediation for Route N, under the base case analysis, is €216,315.

Under the base case analysis, the cost of remediation for Route G is €1,135,941 greater than the remediation cost for Route N. This means it would be cost effective to choose Route N over Route G, unless the construction costs for Route N are at least €1,135,941 greater than construction costs for Route G. While cost differential is an important factor for choosing between alternative routes, there are also other factors that would be taken into account that could change this selection calculus.

10.6.2 Alternative Case: Probabilistic Approach

Our alternative case analysis considers the potential wide-scale adverse impacts to the function and structure of the sensitive alkaline fens ecosystem. These broader

[6]The remediation unit cost used here is purely illustrative. Actual remediation costs for bog forest or wetland remediation are likely to differ and would be dependent on site-specific factors.

scale impacts could occur if road construction in this sensitive habitat type affects overall integrity and function of the habitat unit. Such effects could extend beyond the actual footprint of the roadway. For example, if hydrological changes alter water flows, peat formation, and nutrient cycling, the entire fens could be damaged. Table 10.12 presents three probabilistic scenarios of supplemental ecological harm. All of these scenarios reflect possible ecological effects specific to the construction of Route G, which would pass directly through the alkaline fens habitat. Route N would not pass through the alkaline fens habitat and consequently is assumed to have no impact on the alkaline fens ecosystem.

The first scenario in Table 10.12 assumes that the fens continue to function but that biodiversity losses would extend throughout a broader habitat area (assumed to be 100 ha). Such biodiversity losses could occur from migration barriers or changes in the water table. The probability of this scenario is assumed to be 25%. The total area of affected alkaline fens habitat is 100 ha, with an estimated 40% decline in habitat quality associated with biodiversity impacts. It is assumed that losses do not occur immediately at the time of construction but increase from zero to 40% during a transition period of 30 years. Given the habitat scalar for alkaline fens of 15:1 relative to common marsh (Sect. 10.4) and using the calculations for discounting and normalised debits described above, the total debit from the potential loss of biodiversity is 3,466 DSHaYs. This total loss accounts for the 25% probability that this scenario will occur.

Table 10.12 International road construction—ecological harm scenarios

Alternative scenarios	Scenario probability	Affected area (ha)	Perpetuity ecological service loss relative to full-function alkaline fens	Transitional period (years)	Habitat scalar	Additional normalised debit (DSHaYs)
Remains as alkaline fens, with biodiversity loss	25%	100	40%	30	15	3,466
Becomes common wetland, with full function	25%	100	93%	30	15	8,087
Becomes common wetland, with 20% loss from reduced biodiversity	25%	100	95%	30	15	8,202
Total						19,754

The second scenario assumes that hydrological or other habitat function critical to sustaining alkaline fens are disrupted by the presence of the road in Route G. Because of these alterations in function, the 100 ha of alkaline fens habitat shift to become common marsh over a 30-year transition period. The common marsh is assumed, however, to be fully functioning. An assumed service loss of 93% is calculated based on the habitat scalar of 15:1 for alkaline fens relative to common marsh. The units of loss are then normalised to common marsh, again using the habitat scalar. After discounting, and after accounting for the 25% estimated probability of occurrence, total losses associated with this scenario are 8,087 DSHaYs.

The third scenario assumes ecological losses that would be associated with the transition of alkaline fens to a degraded common marsh (e.g., the common marsh is degraded relative to fully functioning common marsh because of reduced biodiversity). The service assumptions presented in the third scenario result in total ecological service losses of 8,202 DSHaYs. The ecological service flows associated with each of the three ecological risk scenarios are illustrated in Fig. 10.3.

In addition to the three scenarios presented in the table, we assume that there is a 25% probability that no supplemental adverse impacts will occur to the alkaline fens ecosystem. Expected value losses from the first three scenarios therefore represent the total expected losses from ecosystem impacts (i.e., summation of the individual probablistic outcomes). As shown in Table 10.12, the total expected value of debit therefore is 19,754 DSHaYs. Adding these losses to the base case results, total wetland losses associated with Route G would increase from 1,418 to 21,172 DSHaYs. The quantity of required wetland remediation therefore would increase from 81 ha (Sect. 10.6.1) to 1,216 ha. Under the alternative case therefore the total cost of remediation for Route G would be €12,700,000 (again, assuming unit costs of €10,000 for illustrative purposes).

These alternative assumptions regarding potential ecosystem impacts to alkaline fens do not affect the analysis for Route N. The cost of remediation for Route G damages would therefore be €12,500,000 greater than the remediation cost for Route N. This means it would be cost effective to choose Route N over Route G unless the construction costs for Route N are at least €12,500,000 greater than construction costs for Route G. As above, there are of course other factors to consider when choosing between routes.

Finally, we examined an alternative assumption regarding the habitat scalar for alkaline fens. The habitat scalar of 15 was based in part on the scarcity of sedge moss fens relative to total mire habitat in Poland. Alkaline fens is included within the category of sedge moss fens habitat, and sedge moss fens is one of many wetland habitats included within total mire habitat. The ratio of the area of sedge moss fens to total mire habitat is 15. By comparison, however, the area of alkaline fens, by itself, relative to total mire habitat is 1:13,222. If this scarcity ratio is used as the habitat scalar for converting alkaline fens to common marsh, the revised estimate of required wetland remediation would be 1,123,000 ha. The total cost of remediation for environmental damages associated with Route G could be as much as €11 billion.

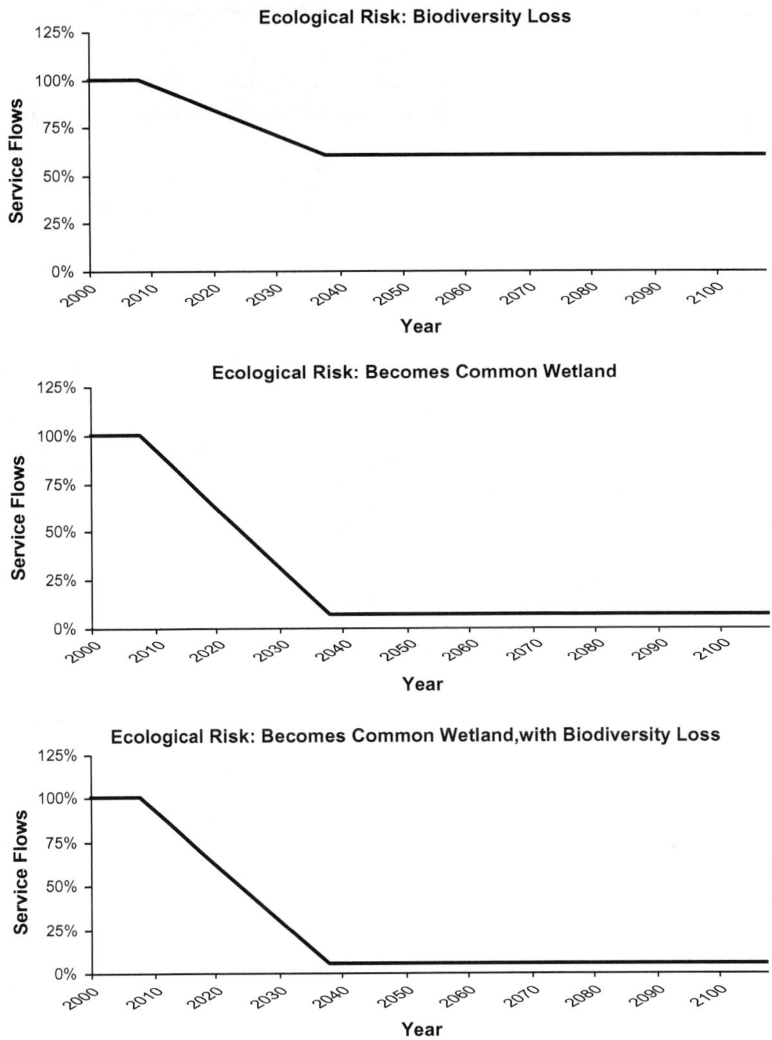

Fig. 10.3 Reductions in ecological service flows associated with disturbance of sensitive alkaline fen ecosystems

10.7 Conclusions

HEA was used to contrast the potential environmental damages associated with two alternative routes of a hypothetical international highway. Application of HEA on an *ex ante* basis enabled us to compare the cost effectiveness of the two alternatives, considering the environmental externalities associated with anticipated future environmental damage. Under a simple base case, environmental damages for Route G were somewhat greater than for Route N. When we considered potential wide-scale ecosystem damages using a probabilistic approach, environmental damages for Route G were considerably greater than for Route N. When this probabilistic approach was expanded further to consider the relative scarcity of the extremely rare alkaline fen habitat that could be lost, environmental damages could increase to over €1 billion.

This case study illustrated how HEA could be applied in an *ex ante* case involving infrastructure development. Further, the case study illustrates application of habitat scalars in resource equivalency. Finally, the case study illustrates a probabilistic approach to estimating expected environmental damages in *ex ante* situations.

Acknowledgements Dr. Aleksandra Sas-Bojarska, Gdańsk University, Poland Eng. Małgorzata Górska and Jarosław Krogulec, Polish Society for the Protection of Birds (OTOP), Warsaw, Poland, for expertise, advice, and support. Additional technical assistance and analysis for this case study was provided by Eng. Andrzej Kamocki, Faculty of Building and Environmental Engineering, Białystok Polytechnic, Poland; Łukasz Meina, The Polish Society for Protection of Birds (PTOP), Białystok, Poland; Dr Wiktor Kotowski, Department of Plant Ecology and Environmental Protection, Institute of Botany, University of Warsaw, Poland. Legal background analyses for the case study was provided by Sandy Luk, then at EPPAC Limited, Letchworth Garden City, United Kingdom. Biodiversity advice with regard to European Union nature conservation policy was provided Zbigniew Karpowicz, Royal Society for the Protection of Birds (RSPB), United Kingdom.

Reference

BBOP and UNEP. (2010, March). *Biodiversity offsets and the mitigation hierarchy: a review of current application in the banking sector*. Business and Biodiversity Offsetts Programme and United Nations Environment Programme.

Chapter 11
Severe Wildfire in a Mediterranean Forest

Roberto Molowny-Horas, Armonía Borrego, Pere Riera and Josep Maria Espelta

Abstract This case study illustrates the equivalency analysis for estimating *ex post* environmental damage and appropriate compensatory remediation following a severe wildfire caused by a power line in a forest protected under the European Union Habitats Directive (HD). The study addresses long-term environmental damage (e.g., over several decades) by a large-scale disturbance in a terrestrial ecosystem, and includes an analysis of uncertainty associated with the potential occurrence of natural future fire events in the area. Accounting for the probability of natural future forest fires directly affects both baseline and compensatory remediation options by reducing the habitat area compared to an assumption of no future forest fires. Only natural forest fires, i.e., 10% of all forest fires, have been included in the calculations of both the baseline and the compensatory remediation, since the operator may not be made liable for accidental or provoked forest fires. The impact of this hypothesis is tested by means of a sensitivity analysis. The case study illustrates:

- Considerations in selecting a metric from various potential ones (hectares, trees, biomass, habitat quality) for terrestrial habitats included in the HD;
- Application of a value equivalency approach (specifically, value-to-value);

R. Molowny-Horas (✉) · J. M. Espelta
CREAF, Cerdanyola del Vallès 08193, Spain
e-mail: roberto@creaf.uab.es

J. M. Espelta
e-mail: josep.espelta@uab.cat

A. Borrego
Center for Research in Environmental Geography, National Autonomous University of Mexico (CIGA—UNAM), Antigua Carretera a Patzcuaro 8701, Morelia, CP, 58190 Michoacan, Mexico
e-mail: aborrego@ciga.unam.mx

P. Riera
Institute of Environmental Sciences and Technology, Autonomous University of Barcelona, ICTA, Edifiici Z, Campus UAB, 08193 Bellatera, Spain
e-mail: prieram@gmail.com

J. Lipton et al. (eds.), *Equivalency Methods for Environmental Liability*,
https://doi.org/10.1007/978-90-481-9812-2_11

- Analysis of key variables (e.g., differences in metrics, single/multiple metrics, on-site/off-site implementation); and
- Sensitivity of the results to changes in four key model parameters (i.e. area of future forest fires, tree mortality, percentage of natural forest fires and tree minimum diameter at breast height).

Keywords Habitat equivalency analysis · Value equivalency analysis Forest wildfire · *Pinus nigra*—simulation model · Spain

11.1 Introduction

This case study illustrates the application of equivalency analysis to the remediation of long-term environmental damage to a terrestrial habitat protected under the Habitats Directive (HD). The so-called *Bages-Berguedà* application (hereafter, BABE) presented here is of interest for two main reasons: first, it shows the application of Habitat Equivalency Analysis (HEA), Resource Equivalency Analysis (REA), and Value Equivalency Analysis (VEA) to the compensation of long-term, large-scale environmental damage that in turn may need large-scale and relatively expensive compensatory measures to be applied. Second, it exemplifies the application of simulation models to deal with the complexity associated with the evaluation of the habitat recovery.

The steps usually followed in applying HEA, REA, and VEA are adapted to suit this particular case study. Section 11.2 describes the incident and the affected habitat and summarises the available data (forest inventories, wildfire datasets, and forest management practices) that will be used in the analysis. Section 11.3 presents the main baseline parameters and the four metrics used in the analysis and discusses the calculation of debit evaluation, including natural, primary recovery, and total interim loss. Section 11.4 describes and assesses the two compensatory remediation options (afforestation of selected areas and fire-prevention plans) that have been selected. In Sect. 11.5, the scaling of the compensatory remediation is calculated, as well as the costs involved, and a sensitivity analysis of key model parameters is presented. Section 11.6 provides a short description of how the recovery of the affected area should be monitored regularly. Finally, Sect. 11.7 summarises the study and discusses the implications of the results.

11.2 Initial Evaluation—the Impact Event

11.2.1 Description of the Incident

Between 4 and 8 July 1994, a large forest fire (Fig. 11.1) occurred in the Catalan counties of Bages and Berguedà, located in northeast Spain (410 45′ to

420 6′ N; 10 38′ to 20 1′ E). The fire was caused by a malfunctioning power line. There are previous cases in the same region and in neighbouring areas where power companies have been declared legally liable for the accidental ignition of wildfires due to poorly maintained power lines (e.g., the Solsonès County wildfire that burned 14,000 ha in 1998). Official reports on forest fires in Mediterranean countries estimate that approximately 17–21% of forest fires are caused by power line malfunctions (Peix i Massip 1999).

The BABE wildfire burned approximately 25,000 ha of European black pine (*Pinus nigra* subsp. *salzmannii*). Black pine forests are included in Annex I of the HD (9530, *Sub-Mediterranean montane forests with endemic black pines*) and they are assigned a high priority for conservation. Sub-Mediterranean black pine forests are present in Italy, Greece, Corsica, and Spain.

The BABE wildfire had an extraordinary impact both in ecological and socioeconomic terms. Black pine is a fire-sensitive species, and natural postfire regeneration is very limited. The new forest landscape that appeared after the fire event included the presence of large areas without any tree regeneration, as well as significant changes in the forest tree species (e.g., mixed oak coppices through resprouting; Retana et al. 2002). Overall the wildfire led to a one-third reduction in the total area of the black pine in Catalonia. The wildfire also impacted popular recreational activities, such as hunting and mushroom-picking, and other tourist-related industries, all of which were drastically reduced after the fire. Black pine does not have any mechanism to survive or to regenerate after intense fire events (Espelta et al. 2003). Therefore, the natural recolonisation of the burned area by black pine is expected to take an extraordinarily long time, or never occur at all, unless proper remediation measures are implemented.

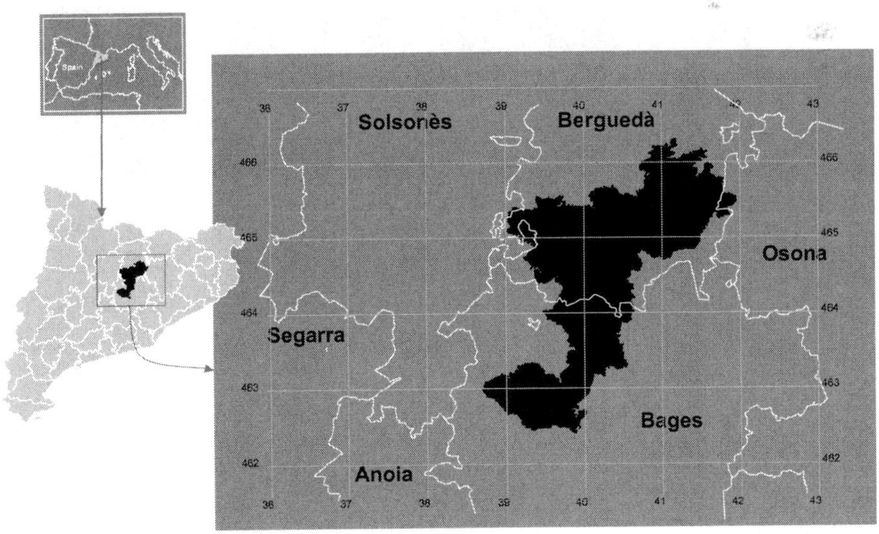

Fig. 11.1 Geographic location of Catalonia and the 1994 BABE fire

11.2.2 Description of the Habitat Affected

The European black pine (*Pino laricio* or *Pino negral* in Spanish; *Pinassa* in Catalan; see Fig. 11.2) is the fourth most abundant tree species in Catalonia in number of stems after *Quercus ilex* (holm oak), *P. sylvestris* (scots pine), and *P. halepensis* (aleppo pine). It is also the fourth most extensively distributed tree species in terms of area (ha) (Burriel et al. 2000–2004). In Catalonia, black pines can be found at elevations ranging from 400 to 1,500 m. Preferring slopes facing north, it can grow up to 40 m or more in height and up to 80 cm or more in diameter at breast height (Bolòs et al. 1993).

Black pine cones open from December to April. Their winged seeds can disperse to nearby areas, even though dispersion distances of up to 100 m or more have been measured in open areas (Ordóñez et al. 2006). If local conditions are appropriate, seeds germinate and establish in the Spring, at the end of which seedlings grow to a few centimeters high.

Black pine lacks any kind of protective or defensive mechanism against fires: its bark is thinner than that of other Mediterranean pine species (like *P. pinaster*); it does not resprout from the stump after a fire (like *Q. ilex*); and it does not have a canopy seed bank (like *P. halepensis*). It is essentially a defenseless species against fires and, as such, it suffers the most during the summer wildfires that regularly affect the Mediterranean forests. Unsurprisingly, its natural regeneration following a severe fire is very low, which together with the high recurrence of forest fires in the Mediterranean basin causes a relentless reduction of its current habitat.

Fig. 11.2 An example of a black pine forest in Catalonia

Black pine forests, referred to as (sub-) Mediterranean pine forests with endemic black pines, are listed in Annex I of the HD and hence can be regarded as a European protected habitat in terms of the Environmental Liability Directive (ELD). The Interpretation Manual of European Union Habitats (European Commission 2007) specifically refers to *Pinus nigra* subspecies *Salzmanni*. The distribution of Salzmann's pine forest is described as follows: *Pinus salzmannii* forests of Spain (Pyrenees, northern Iberian Range, sierra de Gredos, serranía de Cuenca, Maestrazgo, sierras de Cazorla, Segura and Alcaraz, calcareous periphery of the Sierra Nevada) and the Caucasus.

11.2.3 Available Data

There are three main data sources that were used in this analysis which are described in this Section: forest inventory, forest fires, and silviculture practices.

11.2.3.1 Forest Inventory Data

The Spanish 2nd and 3rd Inventarios Forestales Nacionales (National Forest Inventories) were finished in 1990 and 2000–2001 and represent the main datasets with which to carry out the present analysis. Moreover, in 1994, a similar, more ecologically focused endeavor, the Inventari Ecològic i Forestal de Catalunya, i.e., Ecological and Forest Inventory of Catalonia (see Gracia et al. 2002), was completed by several teams from Center for Ecological Research and Forestry Applications (CREAF). Allometric equations relating diameter at breast height and tree height, tree growth, and biomass weight, were computed from this dataset.

In the short time interval from 1990 to 1994 (before the fire), the BABE forest was not affected by events of any type (e.g., fires, road construction) that could have significantly altered its structure or composition. It has been assumed that the 1994 pre-fire BABE forest is basically indistinguishable from the 1990 BABE forest described in the 2nd Spanish inventory. The latter can then be used to describe the structure of the 1994 pre-fire black pine stands.

11.2.3.2 Forest Fire Data

One can reasonably assume that the size and frequency pattern of forest fires in future years will match those of past years. Moreover, as stated in Article 4, Section 1b, of the ELD, the Directive '*shall not cover environmental damage or an imminent threat of such damage caused by a natural phenomenon of exceptional, inevitable and irresistible character.*' It is well known that in Mediterranean-type ecosystems, fires are a common disturbance and may be considered to be part of their natural dynamics (Terradas 1996). If one distinguishes between natural and human-caused forest fires, it must be concluded that operators should be made liable for damages caused by human-induced fires but not by those of natural origin. Previous studies carried out between 1996 and 2005 have shown that in Catalonia, only about 10% of all forest fires are due to a natural cause (i.e., lightning), whereas about 11% are due unknown causes. All other fires are accidental or deliberate (Dirección General de Biodiversidad 2007). Therefore, only natural forest fires were included both in the baseline and in the compensatory remediation calculations.

11.2.3.3 Silvicultural Practices

Traditional management practices consist of selective thinning and pruning. Usually only larger trees are thinned. As a simplification, one can assume that low or moderate cutting does not modify forest density enough as to decrease fire risk. A straightforward, yet realistic, selective thinning approach was implemented whereby trees with a diameter at breast height larger than 30 cm are cut, but only in those stands where basal area is larger than 20 m^2/ha (see Table 11.1).

11.3 Determining the Debits

11.3.1 Baseline Parameters

The temporal evolution of the different metrics was calculated with the aid of appropriate mathematical forest models. Those simulation models assumed that:

- The initial forest was mono-specific and would remain so during the calculations, and
- The forest was almost fully stocked.

These two assumptions considerably simplify the construction of the algorithms and speed up the calculations. As a result of the first hypothesis, the original mono-specific forest does not change its composition in time and no other species need to be introduced in the simulation. The second hypothesis implies that self-thinning, that is, the tendency of less-successful trees dying off as the

Table 11.1 BABE Forest Fire—Forest management practices used in the habitat and resource equivalency analyses

Minimum diameter at breast height to cut (cm)	30
Minimum stand basal area (m^2/ha)	20
Frequency	Annual

Table 11.2 BABE Forest Fire—Variables in the equivalency analysis

Start year	1994
End year	2093
Spatial extent of damage (ha)	25,000
Base year	2007
Annual discount rate (%)	3
Degree of service loss (%)	100
Baseline shape	Dynamic
Recovery rate	Metric-dependent
% of forest fires to affect the BABE area	10

most-successful ones grow bigger, becomes an important factor early in the temporal evolution of the forest.

Table 11.2 provides a summary of the parameters used in HEA/REA calculations. The base year of 2007 was used to calculate all compensatory measures. Consequently, those measures will have to be compounded between 1994 and 2007 and discounted afterward. The BABE area was assumed to have been affected only by 10% of all possible fires, i.e., those of known natural origin.

As described in the 2nd National Forest Inventory, before the fire, there were approximately 25,000 ha of black pine forest in the BABE area, containing approximately 18,150,000 black pine trees. Table 11.3 shows the main structural characteristics of the pre-fire BABE forest.

11.3.2 Metrics

In choosing a measure of loss and gain, or 'metric,' one wants to evaluate the impact of, and recovery from, a given environmental damage in a terrestrial ecosystem. In general, the inner workings of most ecosystems are extremely complex, and their properties and rules are sometimes poorly understood. Ecosystems also interact with their surroundings, which complicates things further. Regarding the BABE forest, one can choose a set of appropriate metrics in order to determine the success of the forest's recovery from the 1994 wildfire. The degree of loss or recovery following implementation of the remediation plans outlined below was determined by calculating the following non-monetary metrics, which are measured per hectare and then summed for the whole BABE area:

Table 11.3 BABE Forest Fire—Structural parameters of the pre-fire BABE forest

Area covered by black pine trees (ha)	25,000
Total number of trees	18,150,000
Tree density (trees/ha)	726
Mean height (m)	8.2
Mean diameter at breast height (cm)	13.6
Mean basal area (m^2/ha)	12.5

- *Total number of trees with diameter at breast height larger than 7.5* cm: A convention was used, as applied in forest inventories elsewhere, such that only trees with Diameter at Breast Height (DBH) (1.3 m) larger than 7.5 cm are counted. The impact of this choice on the final results was tested by means of a sensitivity analysis;
- *Total area covered by trees*: The criteria of minimum canopy cover (>50%) per hectare was used as a threshold to determine whether or not tree cover is high enough to count as a full forest. This assessed the area occupied by the habitat;
- *Total biomass*: Biomass (wood) is directly related to the carbon dioxide (CO_2) captured by the forest, i.e., there is approximately 0.5 kg of carbon in 1 kg of black pine wood, and
- *Total habitat quality index*: A normalised index that took into account the existence of large trees (measured by their DBH and their height) and the total basal area occupied per hectare.

In addition, a welfare or monetary metric is considered to estimate debits and scale remediation using a value-to-value approach. It is measured in monetary terms for the whole affected area, rather than on a per hectare basis, reflecting the welfare loss associated in the interim due to the forest fire and until full restoration back to baseline.

11.3.3 Debits

11.3.3.1 Baseline Determination

A numerical forest model was used in order to assess the complex future evolution of the baseline forest (see Appendix A). The algorithm took into account the non-zero probability of suffering forest fires of natural origin in the future. Fires were subsequently incorporated into the model as a simplified and easy-to-implement stochastic fire submodel (Appendix B), where their annual distributions of size and frequencies were given by the empirical datasets described above. The output of the forest model consisted of the four non-monetary metrics described above and the extent of the burned area per year.

Figures 11.3a, b, c and d[1] show the evolution of the four metrics from 1994 to 2094. To further illustrate the impact of fires, the evolution of the corresponding average metrics with no future fires was also plotted. The plots follow different paths depending on the impact of fires randomly distributed in time and space. Fires cause metric values to differ as compared to a scenario without future fires, and the

[1]The 100-year simulation has been repeated 100 times. Mean 5 and 95% percentile bars are shown in the plots in gray. The baseline without future fires has been plotted in black. The forested area metric remains at a constant value of 25,000 ha because no future fires are included in the calculations.

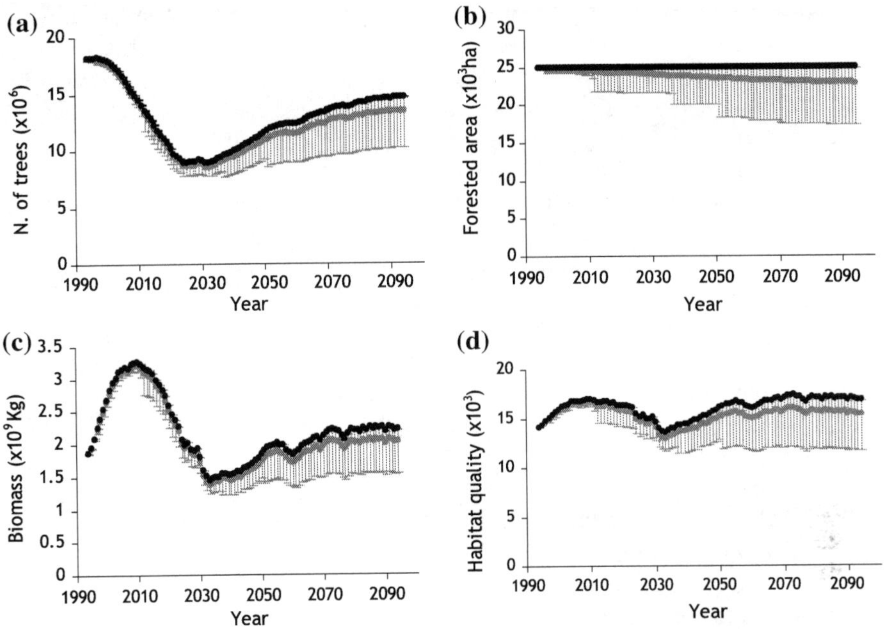

Fig. 11.3 **a** Total number of trees metric, baseline from 1994 to 2094. **b** Total forested area metric, baseline from 1994 to 2094. **c** Total biomass per hectare metric, baseline from 1994 to 2094. **d** Total habitat quality metric, baseline from 1994 to 2094. Metric values without forest fires are plotted in black

different possible trajectories followed by the baseline can be seen as large percentile bars around the mean. Self-thinning also causes the number of trees in the almost fully stocked forest to steadily drop from a starting value of about 18 million to about 8 million and then to slowly increase up to a constant value of about 13 million, which is reached after approximately 150 years (not shown here). The percentile bars are also different in both cases, being undetectable (at the scale of the figure) in the case of no future fires. The impact of future fires is thus twofold: first, they decrease the total number of trees and, second, they introduce a larger scatter in the metric, which stems from the different possible trajectories followed by the simulations. Notice that when future fires are discarded, the value of the forested-area metric remains constant in time, as expected.

11.3.3.2 Natural Recovery

To evaluate the likelihood of the BABE burned area returning by itself to pre-fire conditions, a spatially explicit simulation model of the recruitment of black pine trees from unburned edges was applied (see Molowny-Horas et al. 2007) for a thorough explanation of that model). Seeds from trees along the border of the unburned edge of the forest are dispersed by wind into the burned forest. Some of those seeds may survive, germinate, and then establish and grow. The rate at which this process takes place will determine the speed by which the burned area will naturally recover from the fire.

The evolution of the four metrics as a function of time (not shown here) at several distance intervals from the unburned edge, as simulated with this model, confirmed that the speed of recovery is painfully slow and clearly does not guarantee the recovery of the BABE area within any reasonable time interval. These results were validated by field work carried out in the area (Rodrigo et al. 2004). Moreover, other tree species better adapted to post-fire conditions would soon begin competing with black pine seedlings for light, soil nutrients, and water, further hindering the successful establishment of any black pine trees. It is assumed hereafter that natural recovery is negligible, and consequently primary remediation will be required. For these reasons, only the simulation model of Appendix A is used in the following descriptions.

11.3.3.3 Primary Remediation in the BABE Area: Black Pine Reforestation

As discussed above, the burned BABE area will not return to pre-fire conditions unless steps are taken to remediate the damage. No actions were taken after the fire in 1994 that could count as primary remediation. It is assumed here that to compensate for the habitat loss, the primary remediation strategy to be carried out right after the fire consisted of extensive black pine reforesting in the burned area. Given the extent of the area burned in 1994, tree planting may in practice require several years in order to cover the whole area, depending on budgetary and practical constraints. Moreover, the optimum season to plant black pine seedlings is from February to March, which limits the number of days to plant. As will be explained later, these primary remediation measures could instead be considered as part of the compensatory remediation actions to take place after the fire.

Figures 11.4a, b, c and d[2] show how the proposed primary remediation would return the BABE forest to the baseline. Drawing from the experience of previous reforestation plans in Catalonia, a planting rate of 2,500 ha per year was chosen. Initial density of planted seedlings was 1,500 stems/ha (Espelta et al. 2003). The number of seedlings may have to be increased if initial seedling mortality linked to the early post-transplant period is unacceptably high. The reforestation was assumed to stop after 10 years, when the initial 25,000 ha are replanted (see Table 11.4). However, a percentage of those 25,000 ha was assumed to be lost to new fires. It was also assumed that fires can affect both young plantations and mature trees alike and no effort was made to compensate these losses to 25,000 ha reforested after the initial 10 years.

---

[2]The corresponding baseline metrics from the previous figure are included for reference in black. 5 and 95% percentile bars are plotted around mean values.

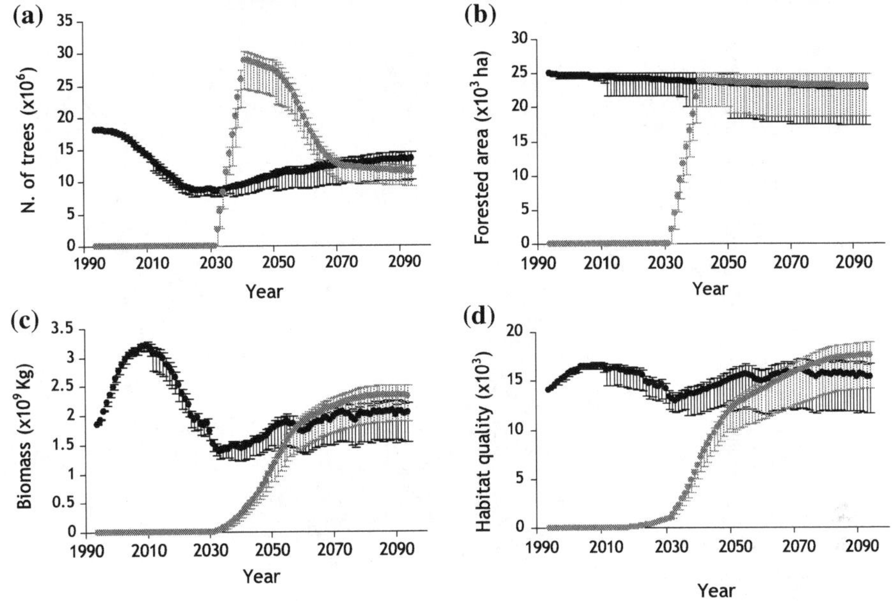

Fig. 11.4 **a** Total number of trees metric, primary remediation (in grey) from 1994 to 2004. **b** Total forested area metric, primary remediation (in grey) from 1994 to 2094. **c** Total biomass per hectare metric, primary remediation (in grey) from 1994 to 2094. **d** Total habitat quality metric, primary remediation (in grey) from 1994 to 2094. Baseline metric values are plotted in black

Table 11.4 BABE Forest Fire Parameters for primary remediation: Black pine reforestation

Start year	1995
End year	2004
Total reforested area (ha)	25,000
Reforested area per year (ha)	2,500
Number of planted seedlings (ha)	1,500

As noted, a 'tree' was defined as a stem with a DBH larger than 7.5 cm. Note that the four metrics in Fig. 11.4 become measurably larger than zero only at the time when the diameters of the earliest plantations become larger than 7.5 cm, which is approximately in 2030, 36 years after the fire. Notice the different behaviours of the four metrics. Although the total reforested area reaches baseline values relatively early (2040), the number of trees very quickly exceeds baseline values (2034) and then slowly converges down to the baseline in the long term. Biomass and total habitat quality, on the other hand, reach values slightly higher than those of the baseline.

11.3.3.4 Determination of Interim Loss

The interim loss to compensate for is defined by the area between the baseline and the primary remediation curves shown in Fig. 11.4. Table 11.5 reflects the

Table 11.5 BABE Forest Fire—Nominal and discounted interim loss calculations

Year	Nominal number of trees loss ($\times 10^6$)	Nominal habitat area loss ($\times 10^3$ ha)	Nominal biomass loss ($\times 10^8$ kg)	Nominal habitat quality loss ($\times 10^3$)	Discount factor (at 3%)	Discounted number of trees loss ($\times 10^6$)	Discounted habitat area loss ($\times 10^3$ ha)	Discounted biomass loss ($\times 10^8$ kg)	Discounted habitat quality loss ($\times 10^3$)
1994	18.14	25.0	18.60	14.10	1.47	26.64	36.71	27.31	20.70
1995	18.06	24.91	19.35	14.30	1.43	25.74	35.52	27.59	20.38
⋮	⋮	⋮	⋮	⋮	⋮	⋮	⋮	⋮	⋮
2033	3.15	19.45	13.45	11.09	0.46	1.46	9.02	6.23	5.1
2034	0.34	17.03	13.39	10.76	0.45	0.15	7.67	6.03	4.8
⋮		⋮	⋮	⋮	⋮		⋮	⋮	⋮
2039		4.79	11.87	8.10	0.39		1.86	4.61	3.15
2040		2.30	11.23	7.39	0.38		0.87	4.23	2.79
⋮			⋮	⋮				⋮	⋮
2056			11.87	2.10	0.23			0.49	1.72
2057			11.23	1.36	0.23			0.31	1.45
⋮				⋮	⋮				⋮
2070				0.06	0.16				0.01
2071				0.04	0.15				0.01
Total						491.37	882.31	966.37	587.93

Note To shorten the table, some of the results were omitted and substituted by the ellipsis. Years for which interim loss becomes negative (and, therefore, will be included as a credit in the calculations) are also omitted

calculation of the interim loss as a function of time for the four metrics. The same table also shows the annually discounted interim loss, with a 3% discount rate. Notice that since the base year for discounting is 2007, values from earlier years are compounded.

The evaluation of interim loss, as shown in Table 11.5, ends when interim loss reaches zero. Nevertheless, compensation from primary remediation may exceed total damage to the BABE area if the number-of-trees metric is used. This is clearly seen in Fig. 11.4, where primary remediation for that metric matches or even exceeds the corresponding baseline. This means there is no need for complementary remediation. Total resource loss will then be overcompensated. That overcompensation will be included in the following calculations as a credit when determining the need for compensatory remediation to offset interim losses. Notice that this would generally not be the case if instead of the number of trees, the metric chosen was the number of equivalent mature trees, for which biomass could be taken as a proxy.

11.3.3.5 Interim Loss in Welfare Terms

The debit estimation in monetary terms was undertaken by a contingent valuation (CV) exercise. The CV method attempts to directly measure the public's welfare loss (debit) by administering a specially designed questionnaire to a sample of the affected human population. The questionnaire for the BABE case study followed the customary approach in natural resource damage assessment: it asked for respondents' willingness to pay (WTP) for a program that would avoid a similar loss to the one that occurred (Adamowicz et al. 1998; Bishop et al. 2000). The loss was defined as the forest damage due to fires measured in terms of hectares of black pine forest (the same metric used in the ecological assessment). While any of the other metrics could also have been used, to have same/similar metrics across different assessment is useful for this case study. The time span of the interim loss until full recovery of the forest was said to be 50 years.

The questionnaire was designed through a series of focus groups and a pilot test during the last quarter of 2007. The full survey took place in the first quarter of 2008. The estimates obtained are considered to be 2008 values. In the main survey fieldwork, a total of 400 individuals were interviewed in person.

The sample was selected within the province of Barcelona, where the fires being assessed occurred. A total of 5.3 million people live in the Barcelona province, of which nearly 4 million are at least 18 years of age. The sample was selected using a mixed approach. The municipalities and the locations within the municipalities were randomly selected according to their population weight. The individuals interviewed within each location followed a gender and age quota representative of the overall population of 18 years of age or older. A typical interview lasted approximately 14 min. No significant problems were detected in the interviewing process.

The willingness to pay question took the form of a ‘single bounded dichotomous choice’, where respondents were asked whether they would be willing to pay € *x* for the proposed program that would avoid the described forest loss. The amount varied in 10 subsamples as: €10, €20, €40, €50, €60, €70, €80, €100, €120, and €150. The payments were to be made every year for 10 years, and they would go up every year according to inflation.

A standard statistical procedure (see, e.g., Hanemann and Kanninen 1999) was used to estimate the WTP distribution, resulting in just under €60 per individual per year to be paid over 10 years. This corresponds to real values of 2008 Euros, since respondents were told that the payment would be revised according to consumer price inflation. A real discount rate could be applied to obtain the discounted values.

11.4 Determining the Credits

11.4.1 Remediation Alternatives

Following primary remediation (tree planting) conducted on site, the BABE forest would be restored to full baseline conditions. This eliminates the need for complementary remediation. However, full recovery will only be achieved after a long time, which brings about the need to introduce compensatory remediation to compensate for the interim losses. The case study evaluated two alternatives:

- An off-site compensatory remediation plan consisting of the afforestation of available areas in Catalonia with black pine trees, and
- A large-scale fire-prevention plan involving splitting large forests up into smaller forested areas separated by firebreaks.

11.4.2 Calculating Credits

11.4.2.1 Off-Site Afforestation

Geographic Information System (GIS) techniques were used to determine the areas in Catalonia (other than BABE burned area) appropriate for planting black pine seedlings. The following set of digital maps were used:

- Map of the potential distribution of black pine in Catalonia (Thuiller et al. 2003), created from forest databases containing climatic and topographic data, in order to determine the areas where black pine trees can grow. These are potential areas in the sense that if seedlings were planted there, they would be able to establish and grow;

- Map of the local slope, calculated with Miramon GIS (Pons 2007) from a 15 × 15 m per pixel digital elevation map elaborated by the Institut Cartogràfic de Catalunya (ICC);
- Land cover map of Catalonia, available from the Generalitat de Catalunya website, and
- Forest fire map of Catalonia from 1975 to 1994 (Díaz-Delgado and Pons 2001).

The areas that meet the following criteria are deemed to be 'appropriate':

- They must correspond to areas not affected by fires at least since 1975, which is the first year for which detailed statistics and digital maps of the extension of fires exist;
- The topography must be such that it is easily accessible by both vehicles and technicians on foot. Following standard practices in Spanish silviculture, an upper limit of 50% was set to the local slope.
- Successful tree planting must only take place where there is relatively little vegetation and where the presence of other tree species is negligible so that large-scale clear-cutting and soil preparation can be avoided. Areas of shrubs, bushes, and abandoned grasslands (which require minimum preparation, e.g., ripping) were selected as potentially appropriate for planting according to digital land cover maps from the Generalitat de Catalunya.
- They must lie within the perimeter of the potential distribution map of Catalonia from 1975 to 1994 (Thuiller et al. 2003).

The Miramon GIS (Pons 2007) spatial analysis tools were used to determine those areas that fulfill the four conditions above. The resulting digital map (Fig. 11.5) illustrates where the compensatory remediation could take place. The map shows a fractioned landscape of small stands scattered mainly over central and north Catalonia, although some appropriate zones do also exist further south. In all, the analysis found 128,000 ha that could be considered to be the most advantageous for the seedling planting scheme to succeed.

Small isolated forest patches will arguably have a lower ecological value than adjacent areas that can eventually merge to create a larger area. The latter effect would reduce fragmentation and in turn increase the ecological value of the forest. To evaluate the amount of fragmentation introduced by the new potential areas determined above, the nearness to existing patches of forest (of any tree species) was taken into account. The results of a GIS analysis of the whole Catalan territory show that the final, i.e., after afforestation, number of forest patches with area smaller than 100 ha (which was chosen as a threshold) will actually decrease, albeit marginally, when the new potential areas are included. That is, afforestation of the scattered areas slightly reduces the fragmentation of the Catalonian forests.

Planted areas can also be affected by fires. The likelihood of such fires was included in the simulation analysis in the same way it was for the primary remediation plan outlined above. Figure 11.4 includes a graphical description of the evolution of the metrics when 25,000 ha are afforested at a rate of 1,500 ha per year.

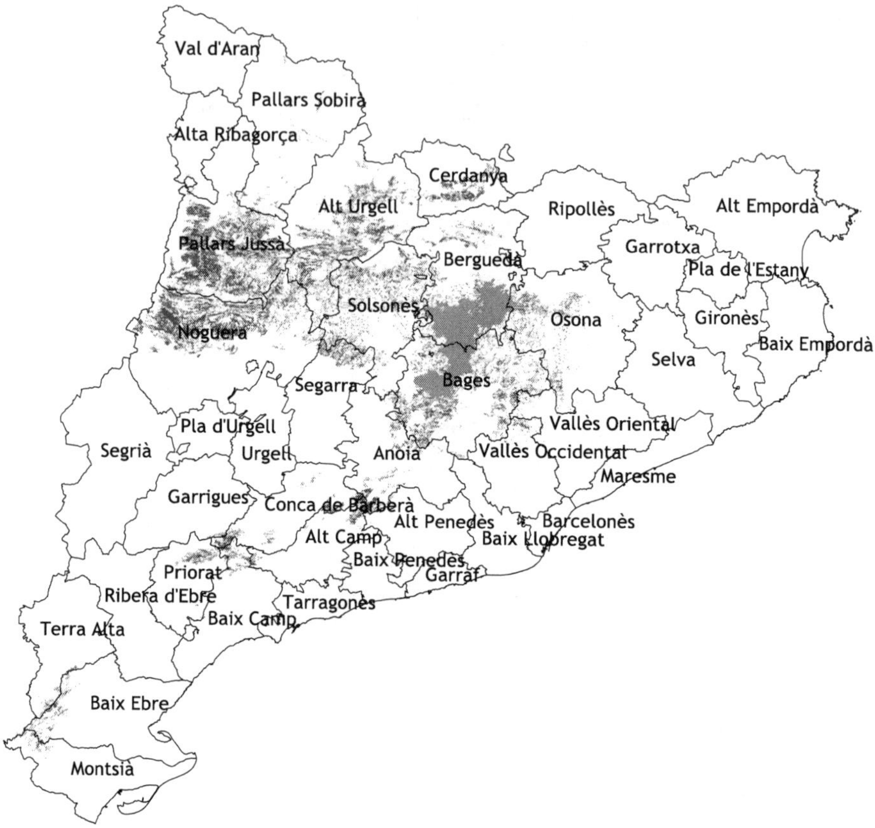

Fig. 11.5 Spatial distribution of optimal zones (gray-shaded areas) for planting black pine seeds. *Note* The continuous gray-shaded area corresponds to the BABE forest, as in Fig. 11.1. The city of Barcelona is located in the 'Barcelonès' county, south-east from the BABE area

The discounted and accrued unit credit per hectare for each of the four non-monetary metrics was calculated for the time period 1994–2093. Results are shown in Table 11.6. These calculations included, as a credit, the overcompensation that stemmed from the implementation of the primary remediation (see above).

This compensatory remediation option was presented in the contingent valuation questionnaire mentioned above. The location of the afforestation area varied across subsamples following the geographical pattern reflected in Fig. 11.5. The size of the afforestation area also varied from 10 to 100% of the BABE damaged surface. Figure 11.6 shows an example of the cards included in the contingent valuation questionnaire.

Table 11.6 BABE Forest Fire—Accrued and discounted unit credit per hectare for the four metrics

Metric	Total unit credit
Number of trees	10,420.6
Forested area (ha)	11.16
Biomass (10^3 kg)	644.26
Total habitat quality index	6.01

Fig. 11.6 Example of location and size of burned area and afforestation area. *Note* The percentage of the BABE burn area (for afforestation), shown in the light-shaded area, used in the credit estimation of the value-to-value exercise. The approximate location and size of the BABE burn area are shown in black

11.4.2.2 Fire Prevention

A fire-prevention plan was implemented in which a forest area was divided into smaller patches separated by firebreaks. These isolated patches prevent a fire that started somewhere within one area from spreading and affecting the whole forest. Instead, the combination of firebreaks and more conventional fire-extinguishing strategies (not evaluated here) limit the size of any possible wildfire.

Instead of simply cutting and clearing swaths of forest to make room for the firebreaks, existing crop fields and other non-forested areas may be used to distribute the different forest patches. This fire-prevention plan involves active management of agricultural land adjacent to forest land. Local authorities or government agencies are assumed to financially support land owners to help them shift crops. These crops should be chosen such that their phenological cycle coincides with the hot season so that they stay green throughout the summer until the fall harvest. Greener crops are much more difficult to burn than dry crops such as postharvest corn and will therefore present a more effective barrier against the propagation of a fire. If managed correctly, these new protective buffer areas will stop fires from progressing from one area to the next. The feasibility of implementing this compensatory measure to the Catalonian forests is not discussed in this analysis, although a similar study for an area adjacent to the BABE forest can be found in Ibáñez et al. (2007). The remediation strategy proposed here is based on the work by those authors.

This particular compensatory remediation strategy can actually be thought of as:

- On-site or off-site *ex ante* (plans can be set up in fire-prone areas to prevent damage on-site or compensate for future fires off-site) or
- Off-site *ex post* (once the fire has occurred, a prevention plan can be set up to compensate for future fires in off-site areas).

In both cases, it is necessary to evaluate the amount of forest that will be prevented from burning every year.

A detailed and exhaustive application of this compensatory strategy to a specific case was beyond the scope of this case study. Instead, a more simplified analysis was undertaken in which an imaginary 5,000-ha forested area was enclosed by a firebreak. The wildfire sub-model described above was then applied to this area to calculate the average amount of forest that was burned per year. Obviously, no single fire could be larger than 5,000 ha in this case. For the sake of illustration, Table 11.7 shows the comparison of average annually burned area between the 5,000-ha forest just described and an unprotected 25,000-ha forest. Scaling the 5,000 ha results up to 25,000 ha yields 160 ha. The difference between averages then gives 46 ha (= 206 − 160) per year.

Table 11.8 also shows the unit credit (discounted and accrued) corresponding to this compensatory remediation strategy for the four metrics of the study. Clearly, the impact of the fire-prevention plan is very small, which becomes apparent when scaling of this compensatory plan is performed (see below).

Table 11.7 BABE Forest Fire—Calculation of average annually burned area for the fire prevention remediation plan

Total area (ha)	Average area burned per year per hectare
5,000 (25,000) Fire prevention plan	32 (160)
25,000 unprotected	206
Difference for 25,000	46

Table 11.8 BABE Forest Fire—Unit credit calculation per hectare for the fire prevention compensatory remediation plan

Metric	Unit credit
Number of trees	61.6
Forested area	0.088
Biomass (10^3 kg)	6.3
Habitat quality index	0.05

11.5 Scaling Remediation

11.5.1 Scaling off-Site Afforestation

The size of the required compensatory remediation in an off-site area from 1994 to 2093 was assessed. Planting rates were similar to those adopted in the primary remediation (see above) and were scaled as needed. Table 11.9 shows the results of scaling off-site afforestation to compensate the interim loss.

The size of the compensatory remediation measure is smaller when the number-of-trees metric is selected, followed by the forested-area metric. Biomass and habitat-quality metrics, on the other hand, which are arguably more related to the actual ecological value of the ecosystem, require much larger compensatory remediation. Table 11.9 clearly indicates that scaling can give different results depending on the metric that is used. It also demonstrates that assessing the ecological implications of the impact of the BABE forest fire requires, as shown here, careful evaluation of several options to carry out an adequate compensation.

As mentioned, a valuation exercise was implemented to estimate the debit or interim loss in monetary units (Sect. 11.3.3). A second contingent valuation exercise was performed to estimate the scale of the compensatory remediation (the

Table 11.9 BABE Forest Fire—Scaling the off-site afforestation remediation plan

Metrics	Size of compensatory remediation (ha)
Number of trees	36,133
Forested area	70,442
Biomass	148,382
Habitat quality index	97,260

credit) that will increase the social welfare (credit) enough to offset the welfare decrease due to the forest fire (debit). Although used to value an environmental improvement instead of avoiding a loss, the questionnaire followed a structure similar to the one used to estimate the debit. The interim loss due to the forest fire was described, and off-site afforestation that would be implemented to compensate for the interim loss was proposed. The cost of this compensatory action would be about €60 per household per year for 10 years, increasing every year with inflation (taken from the first CV used to estimate the debit).

The size of the area where black pines would be planted as compensation varied across the sample (an area of 10% of the damaged site, followed by 20, 30, 50, 70, and 100%). By fixing the monetary bid and varying the environmental change, the minimum size of afforestated area that people require to offset the €60 welfare loss due to the original forest fire was estimated.

In late 2007 and early 2008, the questionnaire went through multiple focus groups and a pilot testing phase. The full survey was implemented in the Spring of 2008. The sample size and selection were similar to that used for the debit exercise. Interviews were also in-person. The mean duration of an interview was 13 min.

Like for the debit estimation, a standard statistical procedure (see, e.g., Hanemann and Kanninen 1999) was used to estimate the necessary amount of compensation in terms of the area of afforestation necessary to compensate for the damage (expressed as a percentage of the originally burned area). The result show that respondents wanted the afforestation area to be 33% of the damaged area, on average.

An interpretation of this result is that under a value equivalency analysis, a forest loss similar to the BABE loss, had it taken place in early 2008, would require a full primary restoration plus an additional 33% of the area damaged as off-site afforestation (for compensatory remediation to address the interim losses). In other words, 1 surface unit of black pine lost due to forest fire requires, in value-to-value terms, 1.33 surface units of black pine afforestation, 1 on-site and 0.33 off-site. Translated into hectares of compensatory remediation, 33% of the original BABE estimated area, which was of 25,000 ha, results in 8,250 ha of off-site afforestation. Compared to the number of hectares needed from the other metrics (Table 11.9), the value-to-value approach requires four times less than the area estimated using the number of trees metric in a resource-to-resource approach.

11.5.2 Scaling Fire Prevention

Scaling for the fire-prevention plan was only calculated for the four non-monetary metrics. The results are shown in Table 11.10. Noticeably, the very large size of the remediation option that would be required for some of the metrics to compensate the interim loss may preclude a direct application of the fire-prevention compensatory remediation. The conclusion is that the fire-prevention compensatory option can be excluded as a valid option for a full compensatory remediation measure.

Table 11.10 BABE Forest Fire—Scaling the fire prevention remediation plan

Metrics	Size of compensatory remediation (10^6 ha)
Number of trees	6.11
Forested area	8.94
Biomass	3.63
Habitat quality index	10.34

11.6 Sensitivity Analysis

A sensitivity analysis of model metrics was performed to understand how uncertainty in some of the input parameters may affect the size of the off-site afforestation remediation plan. Relative sensitivity δC of the size of the compensatory remediation to changes in an input parameter *p*, for a given metric, was defined as

$$\delta C = \frac{\Delta C}{C} \cdot 100$$

where ΔC denotes variations in the area when the input parameter is modified. Changes in the input parameters were introduced one at a time—the parameters were varied $\pm 20\%$ of their nominal value.

Area of forest fires, tree mortality, and percentage of natural forests were selected as the most relevant parameters to be considered in a sensitivity analysis. Relative sensitivity was also measured for the four metrics when the original definition of a 'tree' was modified from 75 to 50 mm DBH. Results are shown in Tables 11.11a, b, c and d.

The results presented in these tables are interpreted as follows. Positive sensitivity implies that the size of the remediation increases when the value of the parameter tested increases. Interestingly, the relative sensitivity to changes in the percentage of natural fires that were included in the calculations is very low. This makes the results presented in this work applicable to cases in which an operator may be liable for a percentage of fires that differs from the nominal value of 10% adopted in this study. The results are also not very sensitive to the variations in the area of future forest fires. This is an indication that this work may also be applicable in future scenarios of higher average temperatures (and therefore higher risk of forest fires) in the Mediterranean area, that can be expected with future climate change. The metric 'number of trees' is the most sensitive of all metrics to a change in the definition of a tree from 75 to 50 mm DBH. This is to be expected since counting trees directly depends on what a 'tree' means. It is no surprise that it is possible to obtain a relative sensitivity close to 40%. The relative sensitivities for the other three metrics are also large. Table 11.11 indicates that including smaller trees in the definition reduces the size of the compensatory remediation plan.

Table 11.11 BABE Forest Fire

(a) Relative sensitivity (%) to changes in the area of future forest fires	
Number of trees	2.1
Forested area	2.0
Biomass	3.9
Habitat quality index	2.4
(b) Relative sensitivity (%) to changes in tree mortality probability	
Number of trees	10.7
Forested area	3.4
Biomass	5.5
Habitat quality index	1.6
(c) Relative sensitivity (%) to changes in the percentage of natural fires	
Number of trees	1.0
Forested area	0.7
Biomass	1.2
Habitat quality index	0.8
(d) Relative sensitivity (%) to changes in minimum DBH of a tree	
Number of trees	40.3
Forested area	24.8
Biomass	6.7
Habitat quality index	6.6

11.7 Cost Estimation

11.7.1 Off-Site Afforestation

Costs were evaluated differently based on whether the areas to afforest were public land or private land. In both cases, costs would be paid for by the operator. In the first case, only the following items were considered:

- Soil preparation of 50 cm depth by ripping with a caterpillar tractor;
- Large-scale purchase of black pine seedlings;
- Transport of seedlings to the sites; and
- Manual seedling planting (including technicians and special equipment).

Costs were estimated using the price lists from a large local reforestation and silviculture company. These numbers should agree well (within reasonable limits) with those from similar companies in Catalonia and elsewhere in Spain. Relatively high contingency costs of 20% are allowed to account for unsuccessful afforestation due to, for example, unfavorable meteorological conditions. The total cost of reforesting/afforesting 1 ha is shown in Table 11.12. All prices provided are in Euros in real values of 2007.

Table 11.12 BABE Forest Fire—Costs of reforesting/afforesting one hectare of public land

Item	Price (€)	Price per hectare (€)
2-year seedling	0.6	900
Soil preparation per km	73	283
Transport and planting	864	864
Total		2,047
Total + 20% contingency		2,456

The costs and funding strategy for the fire prevention plan was assumed to be similar to that of the Common Agricultural Policy of the European Community, whereby owners receive an annual subsidy for 15 years after the afforestation. However, costs were to be paid for by the operator, not by the European Community or any state agency. Those subsidies per hectare were included into the total cost, shown in Table 11.13. A contingency cost of 20% was also included.

A total of 20,000 out of 128,000 ha found in the GIS analysis for suitable areas for afforestation was public land or run by state agencies. Costs for public and private land were then calculated accordingly for each metric. Results are shown in Table 11.14 in Euros in real values of 2007. Prices were computed assuming planting rates were scaled such that afforestation always took 10 years to be completed.

11.7.2 *Fire Prevention*

A full-scale implementation of a fire-prevention plan involved highly complex cost estimations. Given that fire prevention actions generate very low credits, the costs of these actions were not estimated.

Table 11.13 BABE Forest Fire—Costs of reforesting/afforesting one hectare of privately owned land

Item	Price (€)
Planting costs	1,226
Total annual premium (15 years)	2,170
Total	3,396
Total + 20% contingency	4,075

Table 11.14 BABE Forest Fire—Total costs for each metric

Metric	Price (€)
Number of trees	114,871,507
Forested area	254,688,722
Biomass	572,314,258
Habitat quality	363,980,185

11.8 Monitoring and Reporting

The equivalency analyses described here relied on the use of appropriate forest simulation models to determine the type and amount of compensatory remediation action to be provided to compensate for environmental damage following injury. The implementation of simulation models could greatly benefit from better and/or up-to-date datasets from monitoring of how both the burned and the unburned forests change over time. Table 11.15 shows such a monitoring approach. Depending on the type of process to regularly measure or observe, different monitoring time intervals are proposed. Monitoring costs were not estimated. Reports on the recovery trajectory and possible (if any) actions to correct for deviations should be made public every five years.

11.9 Discussion and Conclusions

11.9.1 Estimation of Debit in the Value Equivalency Analysis

In a value equivalency context, the debit could be estimated by benefit (or value) transfer or by a specially designed valuation exercise. Debit and credit valuation could be presented to respondents in the same questionnaire. In this case study, two separate surveys are conducted, first for the debit then for the credit. Given that remediation options will take place over a long term, if the valuation study uses periodic payments of this time period, it is advisable to specify whether payments will vary with inflation or not as future income could vary in any event. As presented here, it was announced that the 10 annual payments would vary according to inflation.

11.9.2 Compensation of Debit

The recommended ecological compensatory remediation plan consists of the off-site afforestation of other suitable areas in Catalonia. Several considerations

Table 11.15 BABE Forest Fire—Monitoring and data collection program for *P. nigra* trees

Item	Time interval
Seedling survival and growth	Yearly
Onset of reproduction	From 20 years of age, every 5 years
Tree diameter growth	Yearly
Changes in fire regime	5 years

could enter the decision-making processes that were not included in this case study, including the availability of enough seedlings and varying market prices.

The conclusion that can be drawn from the analysis here is that metrics used to quantify the damage affect the amount of remediation needed. The most preferred metrics are likely to be total number of trees and total forested area. It is easiest to achieve total compensation using the number-of-trees metric. On the other hand, the amount of compensatory remediation to be implemented for the total forested area will be larger when the other two metrics are considered. This reveals the limitations of using a single metric in describing damage and recovery of the BABE forest.

The value-to-value application from the credit exercise illustrated how the remediation program scaling could be implemented based on a choice contingent valuation approach. With this approach, instead of varying the monetary bid amount, the amount of compensation varies among subsamples. In this way, the mean of the minimum environmental compensation required to offset the welfare loss can be estimated. The metric used in the application was hectares of new black pine forests planted off-site, in a given location, size of which was expressed as the percentage of the damaged area. It was found that as compensatory remediation for interim losses, the damaged forest area was to be compensated by a full primary remediation and by an additional 33% of off-site black pine afforestation in a nearby location.

There is a large difference in the required amount of compensation estimates between the value equivalency and the resource or habitat equivalency applications in this case. The value equivalency approach returns a significantly lower amount of remediation. There can be multiple reasons for such a divergence. One is the usual and, to some extent, inevitable simplification of the ecological change explanation in a questionnaire. This simplification may underestimate the description of damages and thus the compensatory remediation in some instances, even though in some cases the same reason could lead to the opposite situation of overestimation. A second reason can be found on the ecological side. In this case study, the definition of a tree used in the resource equivalency analysis implies that the afforestation renders no credit until almost 40 years after planting. It is likely that younger forests in the on-site restoration and in the off-site afforestation area count for something more than zero in respondents' mind. A third reason is about expectations. There is no need for ecological analyses and social perceptions, on which value equivalency is based, to produce the same results and, indeed, differences could result in either direction.

11.9.3 Natural Versus Accidental or Provoked Forest Fires

Although numbers may vary depending on year and region, approximately 10% of all forest fires in Catalonia have a natural origin (i.e., lightning). The remaining 90% are due to accidents (e.g., sparks from power lines), deliberately started, or are of an unknown cause. In this study only natural forest fires were included both in the baseline and in the compensatory remediation calculations, since the operator cannot be made liable for fires that are not caused by their operations. Moreover, to further check the dependence of the total interim loss on the percentage of natural fires affecting the forest, we performed the same analysis when 100% of all forest fires were included. The result was that although both the baseline and primary remediation differed greatly, the interim loss was less affected by the total amount of fires. That is, fires affect both baseline and primary (and compensatory) remediation in such a way that their effects partially cancel out.

11.9.4 Limitations and Further Improvements

11.9.4.1 Simulation Models

Several simulation models were used to predict the dynamics of a Mediterranean forest. Indeed, predictions about the temporal evolution of forest structural parameters and other ecological indices must be treated with care. In building simulation models, one usually makes a number of assumptions that, on the one hand, simplify the algorithms and make it possible to answer a set of questions within a reasonable time and with relatively limited budgetary and computational resources. On the other hand, those simplifications may give rise to significant errors in mid-term and, especially, long-term predictions. It is therefore advisable to test and validate the simulation algorithms that are used in the study.

11.9.4.2 Inclusion of Smaller Trees

In the present analysis, only stems with diameters at breast height of greater than 7.5 cm were defined as trees and included in the calculations. When that definition was revised to include diameters of 5.0 cm, the results changed, even though only the number-of-trees metric was strongly affected. In general, a scaling factor could be introduced so as to account for smaller stems in some of the other three metrics. Such a factor could in fact be different for different metrics, namely:

- Trees could be weighted by the basal area they occupy, so that smaller stems are still accounted for (although the contribution of smaller stems to total basal area is predictably small);
- Total forested area may take account of the cover provided by smaller stems; better allometric relations between diameter and canopy cover would be required;
- A new habitat-quality metric could be devised to take into account, as a new factor, the potential production of cones depending on tree diameter, since trees of smaller diameter also produce cones, albeit at a much lower rate; and

- The presence or absence of birds could be used as a new metric to quantify the quality of the damaged ecosystems. The wealth of information about birds and their nesting habits in Catalonia (e.g., Atles dels ocells nidificants de Catalunya, 1999–2002) may make this approach possible. Birds may show different nesting and feeding behaviours depending on the size of the trees, which could be factored into this new metric.

As pointed out in the sensitivity analysis, the addition of smaller trees to the calculations drastically changed the number of trees and the forested area metrics, so as to measurably reduce the size of the required remediation option. Biomass and total habitat-quality metrics, on the other hand, were less sensitive to the addition of smaller trees.

11.9.4.3 Datasets

The main datasets used in elaborating the simulation model were the second and third Spanish forest inventories. Finalisation of the fourth Spanish inventory, to be published during the next decade, will make it possible to follow up on the evolution of the different stands during a much longer period of time. The simplified numerical model with which the dynamics and evolution of the damaged BABE forest were explored in the present analysis could then be improved in many ways. One would expect better growth and mortality parameter estimates to be computed from three consecutive forest inventories instead of two, as was done in this study. Diameter growth estimates should also benefit from the availability of new data. These changes should lead to better medium- and long-term predictions about forest structural parameters and, hence, to improved estimates of the metrics.

Further improvements to the predictions put forward by the simulation models may come about from the study of recently planted areas. Growth and mortality of black pine seedlings and juveniles under varying local conditions were calculated in this case study from relatively scarce field and laboratory experiments. Indeed, it would be desirable to develop a careful monitoring and data-collection plan in order to introduce new and better local experimental data from the plantations into the simulation model. These new data, together with improved simulation algorithms, would generate more accurate predictions of the evolution of the metrics.

11.9.4.4 New Species to Be Included

Although the pre-fire BABE forest contained a large proportion of pure or almost-pure black pine stands, there were other tree species with an important presence in the area, which were not incorporated into the analysis above. Those species included, among others, several species of *Pinus* (e.g., *P. sylvestris, P. halepensis*) and of *Quercus* (e.g., *Q. ilex, Q. cerrioides*), which may play a role in shaping the future dynamics of the forest through interspecies competition. The different responses of those trees to competition and water or light limitations will likely determine an evolution of the forest that is different from the one explained here. Nevertheless, the methodology applied to examine the ecological remediation of the 1994 BABE fire will hold true.

Acknowledgements Roberto Molowny-Horas acknowledges the partial financial support of the Spanish Ministerio de Educación y Ciencia and the European Social Fund through the Plan Nacional de Potenciación de Recursos Humanos. Thanks are due to W. Thuiller for providing the authors with maps of potential distribution of *Q. ilex* in Catalonia.

Appendix A: Simulation of the Recruitment of Black Pine Trees in Unburned Forests

Processes affecting tree stands were modeled as deterministic or stochastic functions that in turn depended on stand basal area and/or tree diameter. Those functions determined the temporal evolution of the simulated forest and were empirically determined from the Spanish Inventario Forestal Nacional II (1989–1990) and III (2000) and from the Inventari Ecològic i Forestal de Catalunya. The presence of non-dominant tree species was discarded in the calculations. Consequently, the model behaves as a truly monospecific black pine forest simulation.

The area of study was divided into 1-ha tree plots. The algorithm separately followed the evolution of the cohorts of trees with the same DBH within those tree stands and calculated the four metrics and the area burned per year. Unless so stated, metrics did not consider trees whose diameter at breast height was smaller than 7.5 cm. Climatic variations and their impact on future fires, forest growth, and forest dynamics were not included in the model.

Initially there were numerous parameters to set (typical values are shown in Table 11.16). Model parameters were computed both from field data and from a comparison of 2nd and 3rd forest inventories and field data as follows:

- DBH growth was calculated as a cubic polynomial on DBH for DBH larger than 75 mm. Smaller trees were assumed to reach that DBH in 15 years.
- Tree height and tree canopy diameter were represented by an allometric function, where DBH was the explanatory variable. Parameters were derived from field data.

Table 11.16 BABE forest fire—free parameters in model A

Parameter	Value
Number of simulations	100
Forested area in Catalonia (ha)	1,604,243
BABE area (ha)	25,000
Age of seedlings at a height of 1.3 m	15 years
Discount rate (%)	3%
Number of years to simulate	100

- Forest ingrowth, which is defined as the number of trees that periodically grow into the smallest measured size class of a forest stand, was computed for the following basal areas classes (in square meters): [0, 10), [10, 20), [20, 30), and [30, ∞).
- Tree mortality probability was calculated jointly for each basal area class (such that mortality in denser plots would increase due to competition) and DBH class (such that mortality rate increases when trees are either very small or very large). DBH classes were defined for tree diameters within the following size intervals (in millimeters): [75,125), [125, 225), [225, 425), and [425, ∞).

A single run of the model consisted of 100 time steps, which corresponded to a time interval of 100 years. Initially, a 25,000-ha forest with the same average initial structure (i.e., tree size and age) as the unburned forest was created. The algorithm then proceeded as follows at each time step (see Figure 11.7):

1. Trees in a cohort grow in diameter depending on their DBH.
2. A fraction of those trees may also die depending on their DBH and basal area.
3. Ingrowth takes place depending on basal area.
4. Conditions for forest management practices to be applied are evaluated and implemented if necessary.
5. Forest fires of random size and frequency may affect the simulated forest (see Appendix B).
6. Reforestation/afforestation strategies may be applied if required.
7. Metrics are calculated for trees with DBH larger than 75 mm.

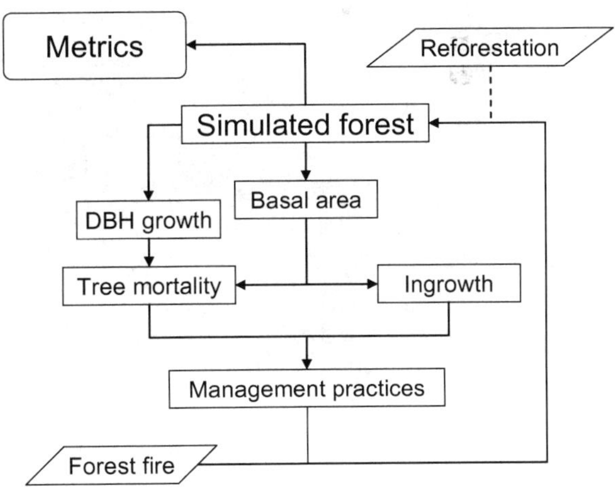

Fig. 11.7 Flowchart of the forest model

Appendix B: Simulation of Future Forest Fires

The three important parameters to consider when modeling fires are:

- Frequency (how many fires have there been per year);
- Size (larger fires will likely be less common than small ones); and
- Location (in general, fires will not be exactly the same as the BABE fire, nor will they burn exactly the same area).

A complete fire propagation model, in which an explicit dependence on meteorological conditions, local topography, and land cover are explicitly introduced, was beyond the scope of this case study. Instead, a simple and very graphic approach to forest fire modeling was adopted. The total forested areas of Catalonia (a total of 1.6 million ha) are analogous to a circular area of 71,459-meter radius, at the center of which is drawn another circular area of 25,000 ha (8,921-meter radius), equivalent to the BABE area (Figure 11.8). Datasets for forest fires were taken from Díaz-Delgado and Pons (2001) for the period 1975–1998 in Catalonia.

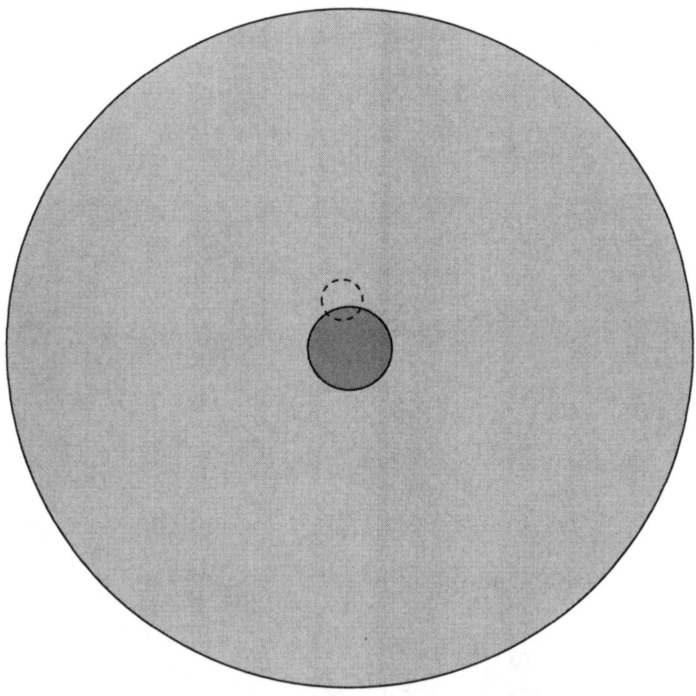

Fig. 11.8 Schematic of the forest area in Catalonia. *Note* The outer circle represents the total forested area of Catalonia. The smaller, darker circle within this, corresponds to the BABE forest. The figure is a proportionate reflection of the burned and total areas. The dashed circle intersecting the central disk corresponds to a 5,770-ha fire, of which 1,300 ha have affected the BABE area in this example

Those data included information about frequency and extent. The stochastic simulation then proceeded as follows: the algorithm picked one year's worth of data at random from the 1975–1998 datasets and distributed those fires at random over the simplified circular area. Each fire from the chosen dataset was also assumed to be circular and may or may not intersect the BABE area. If d is the distance between the center of the two circles, R is the radius of the circular BABE forest, and r is the radius of a circular fire, then the area A of the intersection between the two circles was given by:

$$A = r^2 \cdot \cos^{-1}\left(\frac{d^2 + r^2 - R^2}{2dr}\right) + R^2 \cdot \cos^{-1}\left(\frac{d^2 + R^2 - r^2}{2dR}\right) - \frac{1}{2}\sqrt{(-d + r + R) \cdot (d + r + R) \cdot (d + R - r) \cdot (d + R + r)}$$

If the two circles did not intersect (i.e., the square root is imaginary), it was assumed that the fire did not affect any portion of the BABE area.

References

Adamowicz, W., Louviere, J. & Swait, J. (1998). *Introduction to Attributed-Based Stated Choice Methods*, Final Report to the Resource Valuation Branch of the NOAA Damage Assessment Center, Advanis.

Bishop, R., Breefle, W., Rowe, R., & Wytinch, S. (2000). *Restoration scaling based on total value equivalency: Green Bay natural resource damage assessment final report.* Boulder CO: Stratus Consulting Inc.

Bolòs, O., Vigo, J., Masalles, R. M., & Ninot, J. M. (1993). *Flora Manual dels Països Catalans.* Barcelona: Editorial Pòrtic.

Burriel, J. A., Gracia, C., Ibàñez, J. J., Mata T., Vayreda J. (2000–2004). *Inventari Ecològic i Forestal de Catalunya.* CREAF, Bellaterra. ISBN de la sèrie: 84-931323-0-6.

Díaz-Delgado, R., & Pons, X. (2001). Spatial patterns of forest fires in Catalonia (NE Spain) along the period 1975–1995. Analysis of vegetation recovery after fire. *Forest Ecology and Management, 147,* 67–74.

Dirección General para la Biodiversidad. (2007). *Los Incendios Forestales en España. Decenio 1996–2005.* Ministerio de Medio Ambiente, España ().

Espelta, J. M., Retana, J., & Habrouk, A. (2003). *An economic and ecological multi-criteria evaluation of reforestation methods to recover burned Pinus nigra forests in NE Spain.*

European Commission (2007). *Interpretation manual of European Union habitats EUR 27,* Brussels: European Commission, DG Environment.

Gracia, C., Burriel, J. A., Ibàñez, J. J., Mata, T., & Vayreda, J. (2002). *Inventari Ecològic i Forestal de Catalunya.* CREAF, Bellaterra: Obra completa.

Hanemann, M., & Kanninen, B. B., (1999). The statistical analysis of discrete-response CV data. In Bateman, I. J., & Willis, K. G. (eds.) *Valuing Environmental Preferences. Theory and Practice of the Contingent Valuation Method in the US, EU, and Developing Countries.* Oxford: Oxford University Press.

Ibáñez, J. J., Burriel J. A., Mata, T., Torralba, J., Castro, F. X., & Canyameras, E. (2007). Gestión agrícola y prevención de incendios forestales. *4th Internacional Wildland Fire Conference*, 14–17 May, Sevilla, Spain.

Molowny-Horas, R., Comas, L., Retana, J. (2007). Modelo de simulación forestal de regeneración de *Pinus nigra*. II Reunión del Grupo de Trabajo de Modelización Forestal de la Sociedad Española de Ciencias Forestales, 9–11 May 2007, Jarandilla de la Vera, Spain.

Ordóñez, J. L., Molowny-Horas, R. L., & Retana, J. (2006). A model of the recruitment of *Pinus nigra* from unburned edges after large wildfires. *Ecological Modeling, 197,* 405–417.

Peix i Massip, J., (1999). *Foc Verd II: programa de gestió del risc d'incendi forestall.* Direcció General del Medi Natural, Barcelona.

Pons, X. (2007). *Miramon.* CREAF: Sistema d'Informació Geogràfica I Software de Teledetecció. Bellaterra. ISBN 84-931323-5-7.

Retana, J., Espelta, J. M., Habrouk, A., Ordonez, J. L., & de Sola-Morales, F. (2002). Regeneration patterns of three Mediterranean pines and forest changes after a large wildfire in northeastern Spain. *Ecoscience, 9*(1), 89–97.

Rodrigo, A., Retana, J., & Picó, F. X. (2004). Direct regeneration is not the only response of Mediterranean forests to large fires. *Ecology, 85,* 716–729.

Terradas, J. (1996). El foc en un context mediterrani. Ecologia del Foc. Edicions Proa, Barcelona. J. Terradas (editor).

Thuiller, W., Vayreda, J., Pino, J., Sabate, S., Lavorel, S., & Gracia, C. (2003). Large-scale environmental correlates of forest tree distributions in Catalonia (NE Spain). *Global Ecology and Biogeography, 12,* 313–325.

Chapter 12
Water Abstraction from the River Itchen, Hampshire, United Kingdom

Jonathan Cox and Ece Özdemiroğlu

Abstract The River Itchen is a classic chalk river arising from the chalk aquifer of the Hampshire Downs in central southern England. It is world famous for its fly fishing for trout and Atlantic salmon and was where the techniques of dry fly fishing were first developed in the early 20th century. The river has been used for centuries as a source of power, to irrigate flood plain water meadows and as a source of drinking water. These various uses have had a range of effects on the river and its associated wetlands but despite these many changes it retains a rich biodiversity. This case considers predicted future impacts of abstraction (extraction) for public water supply. This could be an example of 'imminent threat' as defined in the Environmental Liability Directive (Article 2—'sufficient likelihood that environmental damage will occur in the near future'). The removal of water from the river results in reduced water levels and most importantly, reduced flow velocity. This causes a range of effects on the river including increased temperature, reduced oxygen concentration and increased concentration of plant nutrients, particularly phosphate, and other contaminants. Previous investigations have shown that in naturally dry years water abstraction has the potential to cause damage to the populations of Atlantic salmon and the floating *Ranunculus* habitat of the river. This case study uses habitat and resource equivalency analyses to estimate the damage and select compensatory remediation. The economic value of Atlantic salmon is also presented.

Keywords Habitat equivalency analysis · Resource equivalency analysis
Value transfer · Water abstraction · Salmon · Floating *Ranunculus*

J. Cox (✉)
Jonathan Cox Associates, Fig House, Poles Lane, Lymington
Hampshire SO41 8AB, UK
e-mail: jonathan_cox@btconnect.com

E. Özdemiroğlu
eftec, 4 City Road, London EC1Y 2AA, UK
e-mail: ece@eftec.co.uk

J. Lipton et al. (eds.), *Equivalency Methods for Environmental Liability*,
https://doi.org/10.1007/978-90-481-9812-2_12

12.1 Introduction

The River Itchen is a classic chalk river arising from the chalk aquifer of the Hampshire Downs in central southern England. It is world famous for its fly fishing for trout and Atlantic salmon and was the location where dry fly fishing techniques were first developed in the early 20th century. The river has been used for centuries as a source of power, to irrigate floodplain water meadows, and as a source of drinking water. These various uses have had a variety of effects on the river and its associated wetlands. Despite these many changes, the river retains a rich biodiversity.

The river and its associated wetland habitats have been selected as a Natura 2000 site (Special Area of Conservation, SAC) for their representation of the floating *Ranunculus* habitat (listed on Annex I of the European Union Habitats Directive (HD); Fig. 12.1) and for their populations of six species listed in Annex II of the HD, namely:

- Atlantic salmon (*Salmo salar*);
- Bullhead (*Cottus gobio*);
- Brook lamprey (*Lamperta planeri*);
- White-clawed crayfish (*Austropotamobius pallipes*);
- Southern damselfly (*Coenagrion mercuriale*); and
- Otter (*Lutra lutra*).

Fig. 12.1 Floating *Ranunculus* flowering in the River Itchen (copyright Jon Milliken)

The floating *Ranunculus* habitat is characterised by the abundance of water crowfoots *Ranunculus* spp., subgenus *Batrachium* (*R. fluitans*, *R. penicillatus* ssp. *penicillatus*, *R. penicillatus* ssp. *pseudofluitans* and *R. peltatus* and its hybrids). Floating mats of these white-flowered species are characteristic of river channels in early to mid-summer. They may modify water flow, promote fine sediment deposition, and provide shelter and food for fish and invertebrate animals.

Three subtypes of this habitat in the United Kingdom have been described, depending on geology and river type. In each, *Ranunculus* species are associated with a different assemblage of other aquatic plants, such as watercress (*Rorippa nasturtium-aquaticum*), water starworts (*Callitriche* spp.), water parsnips (*Sium latifolium* and *Berula erecta*), water milfoils (*Myriophyllum* spp.), and water forget-me-not (*Myosotis scorpioides*). In some rivers, the cover of these species may exceed that of *Ranunculus* species.

The *Ranunculus* habitat found within the River Itchen provides one of the best examples of subtype 1 in the United Kingdom. Subtype 1 is found on rivers on chalk substrates. The community is characterised by pond water crowfoot (*Ranunculus peltatus*) in spring-fed headwater streams (winterbournes), stream water crowfoot (*R. penicillatus* ssp. *pseudofluitans*) in the middle reaches, and river water crowfoot (*R. fluitans*) in the downstream sections. *Ranunculus* is typically associated in the upper and middle reaches with (*Callitriche obtusangula*) and (*C. platycarpa*).

Water is abstracted from the River Itchen for public water supply at a number of locations in the river's catchment. Seven abstraction licenses have been reviewed by the Environment Agency for England and Wales (EA), with the largest located in the lower Itchen at Twyford, Otterbourne, and Gaters Mill (Fig. 12.2). Water is taken from both the groundwater aquifer and directly from the river (Table 12.1).

Groundwater abstraction at Otterbourne has been shown to have an almost instantaneous impact on river flows due to the close proximity of the wells, adits, and boreholes into the river. Abstraction at Twyford is further away from the river but is likely to have a rapid impact on groundwater flow toward the river.

The HD requires Competent Authorities to review consents considered likely to have a significant effect on Natura 2000 sites. The EA has reviewed consents for water abstraction from the catchment of the River Itchen SAC in accordance with Article 6 of the HD. This has shown that abstraction for public water supply is likely to adversely affect the river's integrity.

For the purposes of this case study, it has been assumed that consents for abstraction for public water supply will be confirmed, despite the negative assessment. As a consequence, compensation would be required to offset adverse effects, in accordance with Article 6(4) of the HD. Alternatively, if the HD did not apply, the anticipated damage due to continued abstraction in the future could be defined as imminent threat under the Environmental Liability Directive (ELD). Article 2 of the ELD defines imminent threat as 'sufficient likelihood that environmental damage will occur in the near future'.

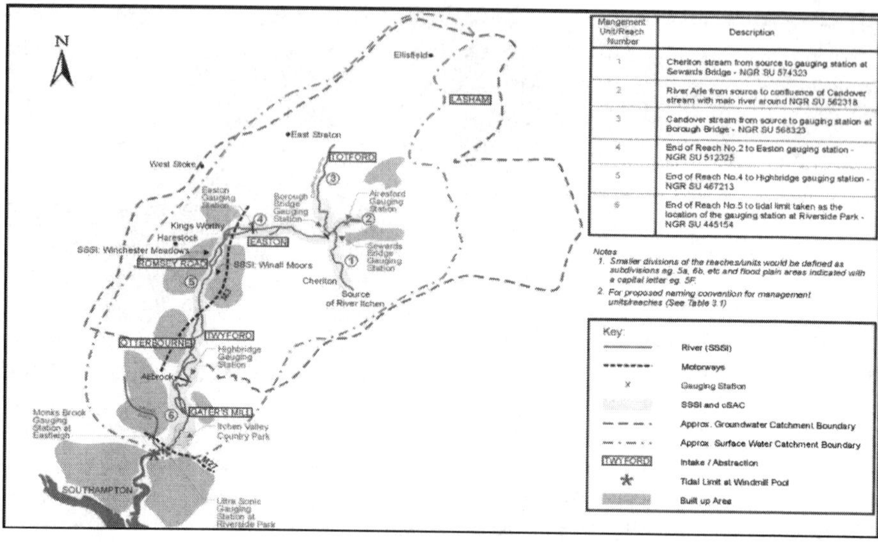

Fig. 12.2 River Itchen catchment showing abstraction points and management units (MUs 1 through 6)

Table 12.1 Water Abstraction, River Itchen—Summary of licensed water abstraction from the catchment for public water supply

	Daily licence (Ml/d)	Annual licence (Ml)
Upper Itchen		
Lasham	27.3	5,455
Totford	4.5	1,659
Easton (Itchen Valley and Winchester)	27.3	6,637
Lower Itchen		
Twyford	36.4	13,320
Otterbourne (including Twyford Moors)	71.6	212,230
Otterbourne surface water	45.5	16,639
Gaters Mill	45.5	16,638

Ml mega liter; *Ml/d* mega liter per day

This short case study demonstrated methods for calculating the magnitude of environmental damage (debit) using two approaches:

- Habitat Equivalency Analysis (HEA) approach using the health of the aquatic macroinvertebrate community as a surrogate for the condition of the floating *Ranunculus* habitat, and
- Resource Equivalency Analysis (REA) approach using predicted numbers of returning Atlantic salmon as a metric.

The economic value of the damage to Atlantic salmon is also shown.

Sensitivity of the debit calculation was investigated using different metrics of invertebrate community structure to measure changes caused by consented maximum water abstraction rates. Credits to compensate for the impact of abstraction were calculated using river restoration works as a chosen remediation method.

The quantum of remediation required was calculated using, both HEA and REA, as done for debit calculation. Differences in the magnitude of compensation estimated through different equivalency approaches to are discussed and compared with economic value of Atlantic salmon (using value transfer of existing evidence —see Chap. 8 for definition of value transfer).

12.2 Initial Evaluation: The Impact

Unlike *ex post* cases considered under the ELD, this case considers predicted future impacts of abstraction for public water supply, as illustrated in Table 12.1. These impacts were not yet observed, as the license holders had not found it necessary to abstract the full volume permitted by their licenses. However, with growing demand for water, it is expected that abstraction quantities will increase in future years. In addition, it is predicted that damage to river biodiversity will become increasingly evident.

The effects of water abstraction on the river ecosystem are complex. Water is taken from the river either directly as surface water or from natural groundwater reservoirs or aquifers. In places, the groundwater abstraction points are immediately adjacent to the river; hence there is hydrological continuity between groundwater and surface water.

The removal of water from the river results in lowered water levels and, most importantly, reduced flow velocity. This causes a range of effects on the river including increased temperature, reduced oxygen concentration, and increased concentration of plant nutrients, particularly phosphate and other contaminants.

The impacts of low flows on the river ecology were investigated as part of the Itchen Sustainability Study (River Itchen Study Group 2004) and subsequently as part of the Review of Consents undertaken by the EA. These investigations have shown that, in naturally dry years, water abstraction has the potential to cause damage to the populations of Atlantic salmon and the river's floating *Ranunculus* habitat.

Impacts to the salmon population will result from reduced numbers of salmon returning from the marine environment, as well as reduced spawning success and survival rates.

Impacts on the habitat were measured by reference to changes in the aquatic macroinvertebrate community. This type of community is typically rich and diverse in chalk rivers and is characterised by a number of species that are dependent on highly oxygenated, swiftly flowing water. Flow thresholds were identified by reference to observed changes in the invertebrate community in high- and low-flow years. The use of invertebrate community data to assess the quality of rivers in general and chalk river habitats in particular was well investigated (Exley 2003; Extence 1981; Extence et al. 1999; Nijboer et al. 2005).

12.3 Determining the Debits

In this section, we consider the baseline situation in the SAC by reference to both the floating *Ranunculus* habitat and the Atlantic salmon population. Because this case study addresses an *ex ante* damage event, baseline conditions are defined as the conditions expected to prevail at the time that full licensed abstractions are initiated. We used current (and recent past) conditions in the river to quantify this baseline. We then considered the impact of full licensed water abstraction on predicted flows in the river. The modelling results were used to estimate the number of salmon that might be expected to fail to return to the river as a consequence of abstraction. Aquatic macroinvertebrate sampling data were analysed in order to identify target flows, below which damage can be expected to occur to the floating *Ranunculus* habitat.

12.3.1 Floating Ranunculus *Habitat in the River Itchen*

Baseline—floating Ranunculus habitat

The river's *Ranunculus* habitat occurs throughout its length and can be assumed as being ubiquitous. However, the condition of the habitat within the river varies and, in some instances, is not in Favourable Conservation Status (FCS). The aquatic macroinvertebrate community present in the river can be considered 'typical species' as defined by Article I of the HD and provide a good indication of the ecological structure and function of the river. As such, they can be used to assess the conservation status of the *Ranunculus* habitat and of the general health of the river (Environment Agency 2004). Analyses of macroinvertebrate survey results related to data on flow provided a powerful tool by which the impact of flow on the aquatic macroinvertebrate community can be predicted and hence act as a surrogate for the conservation status of the habitat.

Summer low flow is a natural feature of the river, and the habitat is able to recover from these natural events[1] (Atkins 2006). However, low-flow events increase in frequency and severity as a consequence of water abstraction for public

[1]Low-flow events occur where flow drops below the long-term Q95 flow (the flow that is exceeded 95% of the time; measured in megalitres/day, or Ml/d). The Q95 is established by creating a flow-frequency curve for the river. Q95 is the flow that is exceeded 95% of the time.

water supply. Using the invertebrate model, a series of low-flow thresholds were set for the six management units in the river. Damage to the protected *Ranunculus* habitat is likely to occur if these are exceeded.

Table 12.2 shows the relationship between long-term flows and target flows for the six management units in the river using the target flow of 0.861 standardised units.[2] Figure 12.2 shows the locations of the management units (MUs); MU1, MU2, and MU3 are all tributaries of the main river. Upper and lower 95% confidence intervals (CI) are also shown.

Abstraction for public water supply that caused flows to fall below the target flow would result in damage to the *Ranunculus* habitat. Due to effects of river augmentation from non-consumptive water users (watercress and fish farms) and because most of the abstraction (83%) takes place in the lower river, the significant effect of abstraction is detectable only in the lower reaches of the river within MU5 and MU6.

To measure the effect of water abstraction on the *Ranunculus* habitat, the extent of the habitat within the river was calculated (Table 12.3). Due to the natural variation in macrophyte cover and composition, this was not considered a good indicator of the extent of the habitat. A better measure involved reference to river flow and bed character. Key flow-dependant invertebrate groups have been described for chalk rivers by Extence et al. (1999), namely, *Baetidae* (mayflies) (Fig. 12.3), *Elmidae* (riffle beetles), *Ephemerellidae* (mayflies), and *Ephemeridae* (mayflies). Invertebrate sampling in relation to habitat has shown that this group of flow-dependant invertebrates is most closely associated with certain river micro-habitats, described as *Ranunculus*, other submerged macrophytes, gravel, and sand.

These flow-dependent habitat types (or micro-habitats) have been used as components of the wider *Ranunculus* habitat for which the SAC has been selected. The EA (Exley 2006) mapped the extent of these micro-habitats in the river. Table 12.4 shows the distribution of *Ranunculus* across the MU5 and MU6 management units, and in total. To provide an area of habitat related to these percentages, the area of each section of river was measured from Geographic Information System (GIS) maps of the designated SAC.

Calculating the debit—floating Ranunculus habitat

Low-flow targets have been set for the two potentially affected sections of the river (MU5 and MU6) based on the invertebrate/flow model. Three targets or rules were established for each management unit. However, for ease of calculation in this case study, only the third rule was used to determine years when adverse effects on the integrity of the River Itchen SAC (damage) is likely to occur, as follows:

---

[2]Standardised flow units were established for the river by relating recorded flows to the long-term mean summer Q95 flow. Flows above the long-term mean scored >1 and flows below the long-term mean summer flow scored <1.

Table 12.2 Water Abstraction, River Itchen—Management unit specific summer Q95 flow thresholds (Ml/d) and target flows

Management unit as in Fig. 12.2	Long-term average summer Q95 (Ml/d)	0.861 Target		
		Lower confidence limit 0.719	Mean 0.861	Upper confidence limit 0.951
1	27.6	19.8	23.8	26.2
2	96.6	69.4	83.2	91.9
3	26.7	19.2	23.0	25.4
4	256.2	181.8	217.9	240.7
5	275.4	197.9	237.3	262.0
6	270.2	194.2	232.8	257.1

Table 12.3 Water Abstraction, River Itchen—Percentage cover of *Ranunculus* habitat per management unit of the River Itchen

	MU5 Main River	MU5 Navigation	MU6
Cover of *Ranunculus* habitat (%)	3	8	3

Fig. 12.3 Larvae of mayfly (*Baetidae*) (Courtesy of Kevin Exley, Environment Agency)

Table 12.4 Water Abstraction, River Itchen—Distribution of *Ranunculus* habitat within MU5 and MU6

	Total area of River (ha)	Percentage of *Ranunculus* habitat	Area of *Ranunculus* habitat (ha)
MU5	35.82	77.9%	27.90
MU5 navigation	10.37	77.5%	8.04
MU6	24.47	72.5%	17.74
Total area of habitat			**53.68**

Management Unit 5
Flow should not fall below 198 ML/d for any period of time
Management Unit 6
Flow should not fall below 194 ML/d for any period of time

Data were obtained for a 20-year period during which flows were monitored within the two MUs of the river that are threatened by water abstraction. Years in which low flows are predicted to exceed the target levels were identified within this period. For both MUs, the same years caused the target flow to be exceeded, providing a pattern of eight low-flow years during the 20-year period, as shown in Fig. 12.4.

This pattern was then projected forward to predict potential low-flow patterns over the next 20 years using 2008 as the base year. There are clearly a number of assumptions in this approach, perhaps most importantly, no account was taken of potential changes in the frequency of low-flow years due to climate change. To take these additional factors into consideration in the prediction of future low-flow events is beyond the scope of this case study. However, if such an equivalency analysis were to be performed in an actual situation, Competent Authorities may wish to consider future environmental states under climate change scenarios.

To estimate the magnitude of damage to the invertebrate community (and by implication the habitat) during low-flow years, comparison was made between four sets of variables (Exley 2004):

- Number of taxa (richness);
- Evenness (measured using the Shannon-Weaver index);
- Total invertebrate abundance; and
- Abundance of flow-dependent invertebrate groups in MU5 and MU6.

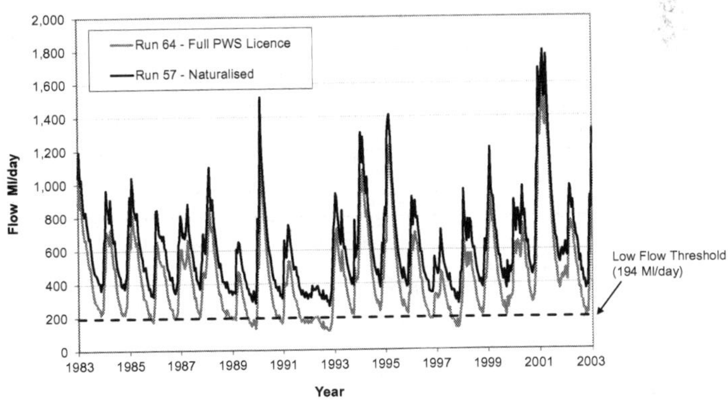

Fig. 12.4 Hydrograph for the lower River Itchen (1983–2002) for MU6 showing the frequency of modelled, naturalised low-flow years and the effects of public water supply licenses on breaching the low-flow threshold

Significant flow-related responses of the invertebrate community do not occur progressively with declining flow. However, they have been shown to occur only below a threshold flow band. All four variables were subjected to analysis of variance (ANOVA) tests and showed significant differences between above- and below-flow threshold years, as shown in Table 12.5. More details of the identification of the threshold flow are given in Appendix to this chapter.

ANOVA: analysis of variance

The percent change in both richness and evenness is of similar magnitude, while the abundance of individuals (both flow dependent and total invertebrate abundance) shows a much greater effect. To illustrate the effect of these two measures on the total damage and hence compensation requirement, an analysis was made using both the change in number of taxa (12.5%) and the change in the abundance of key invertebrate species (71%).

Having calculated the total area of habitat in the damaged sections of the river, the years when damage is predicted to occur (by projecting frequency and pattern of low-flow years from historic hydrograph), and the magnitude of the damage that occurs in low-flow years, it is possible to calculate the damage caused to the habitat each year. For the purposes of this case study, the two most extreme rates were used to calculate the annual loss of habitat service in low-flow years (12.5 and 71%). Other rates of change in the range, shown in Table 12.5, could also have been used and a mean taken. However, the change in flow-dependent species was considered

Table 12.5 Water Abstraction, River Itchen—Results of ANOVA tests comparing the diversity (richness and evenness) and total abundance of invertebrates collected in samples above and below the flow threshold

Criterion	Mean in samples collected above flow threshold	Mean in samples collected below flow threshold	ANOVA Result, p value	Change	% Change
Richness (number of taxa)	40	35	<0.001	Significant decrease	12.5
Evenness	0.62	0.73	0.001	Significant increase	17.7
Total invertebrate abundance	4,549	1,024	0.001	Significant decrease	77.5
Abundance of flow-dependent invertebrates	414.2	121.1	0.001	Significant decrease	70.9

Note The natural question about this comparison is, how much is above and how much is below the low-flow thresholds? Ideally, one would think about a continuous scale: sufficient flow would equal 100% of invertebrate services. Wholly insufficient flow (dry, or close to it) would yield 0% service. There would then be a continuous relationship (maybe concave and exponential) where reduced flow would be mapped against invertebrate impairment. This could not be estimated for this case study, which used the above simplified relationship most likely to reflect changes to the *Ranunculus* habitat. As noted above, the strict 'threshold' concept probably is an oversimplification. Although this is reasonable for a simple case study, it is unlikely to be defensible in a full ELD implementation.

Because the predicted damage to the river will continue for an indefinite period into the future, a period of 100 years has been used over which to calculate damages, with a 3% discount rate to express the changes over time in present value terms (Discounted Service Hectare Years, DSHaYs).

The calculations showing the total DSHaYs over 100 years using both a 12.5 and 71% annual rate of damage are shown in Tables 12.6 and 12.7.

It was assumed that the habitat will recover in one year after a low-flow year, providing flows return to above-threshold levels. However, when there are a series of low-flow years (as between 2013 and 2016), there is no recovery between years and hence the damage is compounded over this time. It might be expected that the rate of recovery would be longer than one year following a series of damaging low-flow years. However, the data available did not appear to support this prediction. If data were available, it would be possible to develop a more complex modelling approach that uses different recovery rates for different degrees of flow reduction and to consider multiyear conditions to identify any increased levels of damage following a series of low-flow years.

Comparison of annual damage rates—floating Ranunculus habitat

The change in abundance of key invertebrate groups of 71%, which was used to calculate annual service losses in Table 12.6 and Fig. 12.5, gave a total habitat-service loss over 100 years of 623 ha of floating *Ranunculus* habitat. By comparison, the use of the change in species diversity of 12.5%, shown in Table 12.7 and Fig. 12.6, gives a habitat service loss over the same period of only 165 ha of floating *Ranunculus* habitat.

This raises the obvious question of which of these two damage rates most accurately reflects the impact of reduced river flow, caused by abstraction for public water supply, on the protected habitat of the River Itchen. Ecologically, it might be assumed that the macroinvertebrate fauna is adapted to low river flows, as is demonstrated by the rapid rate of recovery after low-flow years. Low flow, therefore, has a limited impact on species diversity because most species survive the low-flow events in localised sections of the river or patches of river bed where flow conditions remain tolerable. However, changes in abundance of the key flow-dependent invertebrate groups reflect more accurately the change in extent of suitable habitat within the river during these low-flow events. Also, it is considered a better measure of the impact of water abstraction on the condition of the protected riverine habitat in this case. Note that abundance is typically a less sensitive indicator of contaminant effect.

Table 12.6 Water Abstraction, River Itchen—Debit calculation: 71% annual service loss, change in abundance of key chalk river macroinvertebrate species, years in which the minimum flow threshold is exceeded

Low-flow year	Year	Service years available pre abstraction (ha)	Service years available post abstraction (ha)	Annual service loss (%)	Year, base year	Discount factor	Loss	Discounted loss (DSHaYs)
(A)	(B)	(C)	(D) = (C) × (1 − (E))	(E) = 71.0%	(F)	(G) = 3.00%	(H) = (C) − (D)	(I) = (G) × (H)
No	2008	53.68	53.68	0.00	0	1.00	0.00	0.00
No	2009	53.68	53.68	0.00	1	0.97	0.00	0.00
Yes	2010	53.68	15.57	71.00	2	0.94	38.11	35.92
Yes	2011	53.68	4.51	91.59	3	0.92	49.17	44.99
No	2012	53.68	53.68	0.00	4	0.89	0.00	0.00
No	2013	53.68	53.68	0.00	5	0.86	0.00	0.00
Yes	2014	53.68	15.57	71.00	6	0.84	38.11	31.92
Yes	2015	53.68	4.51	91.59	7	0.81	49.17	39.98
Yes	2016	53.68	1.31	97.56	8	0.79	52.37	41.34
Yes	2017	53.68	0.38	99.29	9	0.77	53.30	40.85
No	2018	53.68	53.68	0.00	10	0.74	0.00	0.00
No	2019	53.68	53.68	0.00	11	0.72	0.00	0.00
No	2020	53.68	53.68	0.00	12	0.70	0.00	0.00
Yes	2021	53.68	15.57	71.00	13	0.68	38.11	25.95
Yes	2022	53.68	4.51	91.59	14	0.66	49.17	32.50
⋮	⋮	⋮	⋮	⋮	⋮	⋮	⋮	⋮
No	2107	53.68	53.68	0.00	99	0.05	0.00	0.00
No	2108	53.68	53.68	0.00	100	0.05	0.00	0.00

Notes
53.68: 100% service provision in hectares in the baseline
3.0%: Discount rate
71%: Annual percent loss of services (based on decline in abundance of key macroinvertebrate groups)
2008: Base year (the year the analysis occurs in)
DSHaYs: Discounted Service Hectare Years
To shorten the table, some of the results were omitted and substituted by the ellipsis

Table 12.7 Water Abstraction, River Itchen—Debit calculation: 12.5% annual service loss, change in macroinvertebrate species diversity, years in which the minimum flow threshold is exceeded

Low-flow year	Year	Service years available pre abstraction (ha)	Service years available post abstraction (ha)	Annual service loss (%)	Year, base year	Discount factor	Loss	Discounted loss, (DSHaYs)
(A)	(B)	(C)	(D) = (C) × (1 − (E))	(E) = 12.5%	(F)	(G) = 3.00%	(H) = (C) − (D)	(I) = (G) × (H)
No	2008	53.68	53.68	0.00	0	1.00	0.00	0.00
No	2009	53.68	53.68	0.00	1	0.97	0.00	0.00
Yes	2010	53.68	46.97	12.50	2	0.94	6.71	6.32
Yes	2011	53.68	41.10	23.44	3	0.92	12.58	11.51
No	2012	53.68	53.68	0.00	4	0.89	0.00	0.00
No	2013	53.68	53.68	0.00	5	0.86	0.00	0.00
Yes	2014	53.68	46.97	12.50	6	0.84	6.71	5.62
Yes	2015	53.68	41.10	23.44	7	0.81	12.58	10.23
Yes	2016	53.68	35.96	33.01	8	0.79	17.72	13.99
Yes	2017	53.68	31.47	41.38	9	0.77	22.21	17.03
No	2018	53.68	53.68	0.00	10	0.74	0.00	0.00
No	2019	53.68	53.68	0.00	11	0.72	0.00	0.00
No	2020	53.68	53.68	0.00	12	0.70	0.00	0.00
Yes	2021	53.68	46.97	12.50	13	0.68	6.71	4.57
Yes	2022	53.68	41.10	23.44	14	0.66	12.58	8.32
⋮	⋮	⋮	⋮	⋮	⋮	⋮	⋮	⋮
No	2107	53.68	53.68	0.00	99	0.05	0.00	0.00
No	2108	53.68	53.68	0.00	100	0.05	0.00	0.00
								164.79

Notes
53.68: 100% service provision in hectares prior to impact (Sect. 3.2.1)
3.0%: Discount rate
12.5%: Annual percent loss of services
2008: Base year (the year the analysis occured in)
DSHaYs: Discounted Service Hectare Years
To shorten the table, some of the results were omitted and substituted by the ellipsis

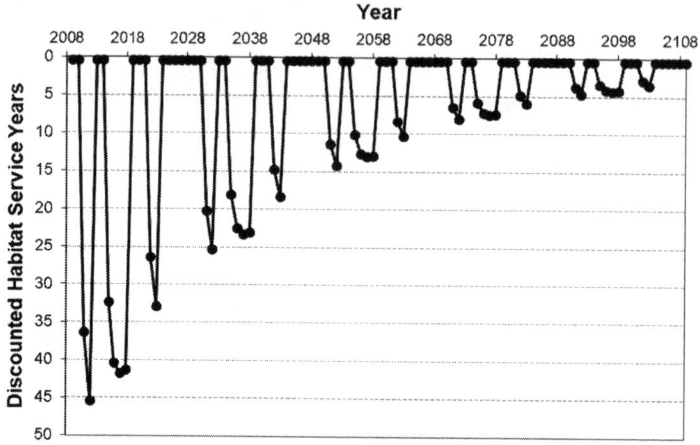

Fig. 12.5 Present value habitat service loss over 100-year period and a 71% annual habitat service loss showing the influence of discounting

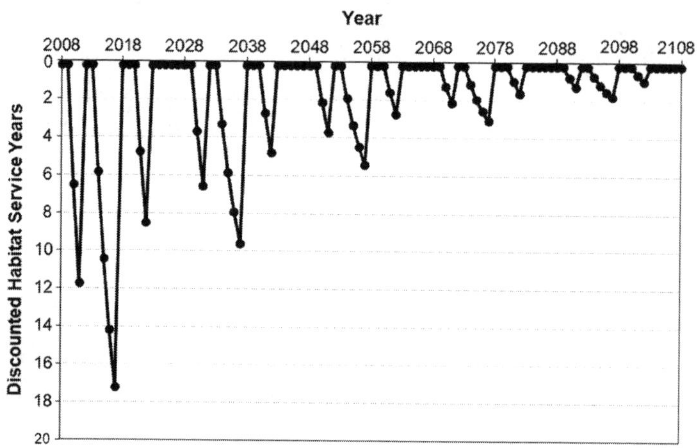

Fig. 12.6 Present value habitat service loss over 100-year period and a 12.5% annual habitat service loss showing the influence of discounting

12.3.2 The Atlantic Salmon in the River Itchen

Baseline—the Atlantic salmon

Atlantic salmon have been a feature of the River Itchen since the last ice age. It is likely that they contributed to the siting of early settlements in the area of Winchester, because prior to agricultural development, salmon were a good source of protein in the winter months. Private rights of net fishing were granted by the

King before Magna Carta, and many documents exist in the Hampshire Records Office of leases of salmon fishing rights by the Bishop of Winchester from the 16th century onward (Solomon 2002).

Taking into account a number of factors, the EA has calculated the egg deposition and consequently the approximate minimum number of adult salmon required for a self-sustaining population in the River Itchen. This 'conservation limit' equates to approximately 660 spawning salmon, or three and a half times the spawning escapement observed between 1999 and 2001. This low population size is thought to be due to several important factors including poor egg survival and poor marine survival.

Several studies have shown that spawning gravel areas of the River Itchen are in poor condition (Scott and Beaumont 1993; Riley et al. 1998; Solomon 2004), with egg survival rates often less than 5%. Riley et al. demonstrated that mitigation methods such as channel modification, gravel reinstatement, and gravel cleaning can increase egg survival.

Identifying a baseline Atlantic salmon population for the River Itchen is problematic. Evidence from the 1990s suggests a declining population. However, more recent data for the period 2001–2006 suggest something of a recovery in population, with approximately 400 returning fish, as illustrated in Fig. 12.7. If this recovery is sustained, it is possible that the population can be restored to a favourable condition.

To simplify this case study, it was assumed that the numbers of salmon returning to the river from the Woodmill Pool equate to the numbers spawning—this is something of an oversimplification because a degree of mortality is to be expected between entering the river and spawning. It was also assumed that the conservation

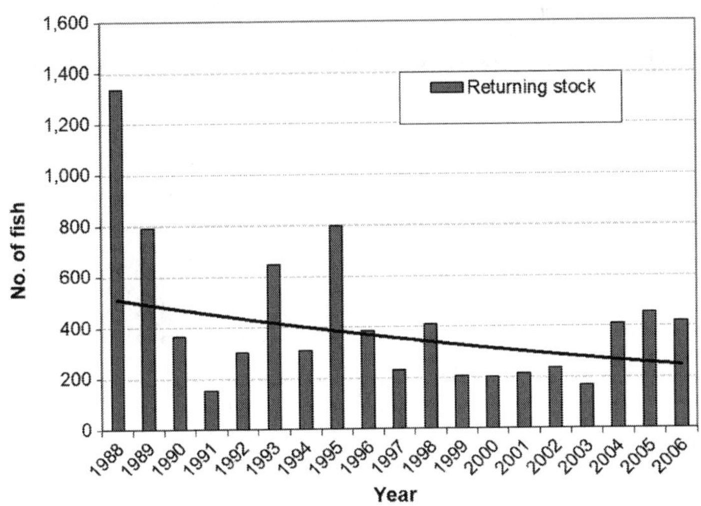

Fig. 12.7 River Itchen returning Atlantic salmon (1988–2006) (Environment Agency and CEFAS 2006)

limit of 660 spawning fish represents the minimum number to achieve FCS, as defined by Article 1 of the HD. Once this population has been reached, Competent Authorities could assume that compliance with Natura 2000 had been achieved. However, it could also be argued that FCS is not reached until a theoretical carrying capacity for the river has been reached. This could be based on reconstruction of historic population size from rod catch returns or on habitat-quality assessments.

The number of salmon returning to the river showed signs of recovery, rising to 419 fish in 2006. It was not possible to determine if this trend was sustainable and likely to continue. However, for the purposes of this case study, it was assumed that recovery will continue at approximately the rate of 6% per annum seen between 2001 and 2006.

To illustrate the effect of choosing different rates of recovery, a 2.5% recovery rate was also used for comparison. This was based on estimates of smolt survival published in United States literature.

It was assumed that there is no longer an adverse effect on site integrity once the baseline reaches the conservation limit for the river of 660 fish. The baseline was therefore considered to be recovering until the conservation limit equating to FCS is reached. In reality, it is hoped and presumed that salmon populations will continue to increase beyond this level. However, these additional fish should not be subject to further remediation once the population has been restored to FCS.

A number of models were developed to predict the impact of water abstraction on the River Itchen Atlantic salmon population. The model that provided the best indication of the effect of abstraction on the numbers of salmon returning to the river was the salmon migration model. This model was based on work undertaken by the EA (2006) and Fewings (2004) on salmon migration related to river flow. It was based on the premise that salmon require certain flow characteristics in order to return from the estuary to the river. Salmon that remain in the estuary for longer periods due to inadequate river flow are vulnerable to fishing activity and natural predation.

The migration model was used to predict the effects of different scenarios on the number of salmon returning past the tidal limit during a high-flow year (2000), average-flow year (1987), and low-flow year (1992), as illustrated in Table 12.8.

Dry years with full licensed abstraction result in a 48.5% reduction in the numbers of salmon returning to the river at Woodmill Pool. Unremarkable years result in 11.3% reduction in numbers of returning salmon, while in wet years there

Table 12.8 Water Abstraction, River Itchen—Loss of salmon (% of run returning to Woodmill Pool) due to two abstraction scenarios compared to naturalised flows

	Naturalised (scenario 9)	Contemporary (scenario 1)	Full entitlement (scenario 10)
Wet year (2000)	0	1.4	3.5
Unremarkable (near-average) year (1987)	0	4.2	11.3
Dry year (1992)	0	30.9	48.5

Note As per Table 12.5, a simplifying set of assumption is used here

would be a 3.5% reduction. The number returning to the river at Woodmill is not necessarily the same as the number of spawning salmon because a further reduction in salmon numbers can be predicted in the river due to mortality. However, for the purposes of this case study, the number of salmon returning to the river at the tidal limit (Woodmill Pool) was taken as equivalent to the number of spawning salmon.

Calculating the debit—the Atlantic salmon

The salmon migration model was used to calculate the percentage of salmon unable to return to the river under high-, average-, and low-flow years, as shown in Table 12.5. Hydrological data from the 20-year period 1983–2002 were used to identify the number of years it might be reasonable to expect these three levels of flow. This is illustrated in Fig. 12.8. The results from the analysis in Fig. 12.8 are shown in Table 12.9, which assigns each year to a high-, average-, or low-flow category.

Comparison of recovery rates—the Atlantic salmon

Table 12.10 shows the results of the debit calculation using a 6% recovery rate to baseline. It gives a total of 4142 Discounted Atlantic Salmon Service Years (DASSYs) lost over a 100-year period. By comparison, using a 2.5% recovery rate, losses are reduced to 3,841 DASSYs over the same period (Table 12.11). This is not a significant difference and reflects the assumptions about recovery back to baseline of 660 fish and discounting. The more the attenuated recovery rate scenario generates losses further in the future, the less is the difference between the two scenarios.

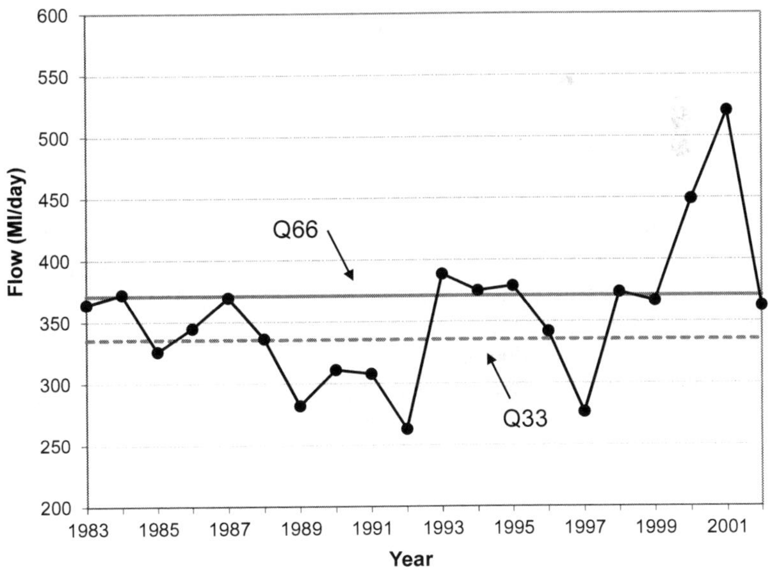

Fig. 12.8 River Itchen minimum annual flows 1983–2002 showing Q33 and Q66 flow thresholds used to identify high-, average-, and low-flow years

Table 12.9 Water Abstraction, River Itchen—Allocation of years to flow category

Year	Minimum recorded flow (Ml/d)	Flow category
1983	364.3	Average
1984	372.4	High
1985	325.9	Low
1986	344.7	Average
1987	369.6	Average
1988	336.2	Average
1989	282.0	Low
1990	311.0	Low
1991	307.8	Low
1992	263.0	Low
1993	388.6	High
1994	375.1	High
1995	379.0	High
1996	341.9	Average
1997	276.6	Low
1998	373.8	High
1999	366.7	Average
2000	449.1	High
2001	520.5	High
2002	362.4	Average

In both of the above calculations, a long time period of 100 years was used to calculate debits. This was done so that credit, in terms of compensatory habitat, can be calculated to maintain the integrity of the Natura 2000 network in perpetuity. Figure 12.9 illustrates the reduction in annual loss over time using discounting and the two recovery rates.

Estimating the debit in monetary terms—the Atlantic salmon

The aim of this section is to estimate the economic cost of the decline in the population of Atlantic salmon in monetary terms and to illustrate the value-to-value and value-to-cost approaches. The economic cost is calculated as the discounted sum of annual economic loss, which is, in turn, the number of salmon lost multiplied by the economic value of one salmon. While economic value could include both market and non-market components (see Chap. 8), the intention is not for the responsible party to make monetary compensatory payments to the affected parties for commercial (market) loss. The principle that money exchange in the context of the ELD must be to compensate the damage resources and their services is retained, even if the metric used to measure damage and remediation is money.

Various economic valuation methods can be used to obtain a unit value for salmon in the River Itchen. One can either undertake a valuation study at the River Itchen site or use previous estimates from the available literature. For the purposes of this case study, we implemented the second approach, which is called value

Table 12.10 Water Abstraction, River Itchen—Debit calculation: 6% annual recovery, changes in Atlantic Salmon, years 2017–2106

Year	Recovery to baseline, number of fish	Baseline +	Fish lost due to water abstraction (%)	Number of fish lost	Number of fish returning to river	Year, base year	Discount factor	DASSYs
	(A)	(B) = 6%	(C)	(D) = (B) × (C)	(E) = (B) − (D)	(F)	(G) = 3%	(H) = (D) × (G)
2008		400	11.30	45.20	354.80	0	1.00	45.20
2009	24.00	424.00	3.50	14.84	409.16	1	0.97	14.41
2010	25.44	449.44	48.50	217.98	231.46	2	0.94	205.47
2011	26.97	476.41	11.30	53.83	422.57	3	0.92	49.27
2012	28.58	504.99	11.30	57.06	447.93	4	0.89	50.70
2013	30.30	535.29	11.30	60.49	474.80	5	0.86	52.18
2014	32.12	567.41	48.50	275.19	292.21	6	0.84	230.47
2015	34.04	601.45	48.50	291.70	309.75	7	0.81	237.18
2016	36.09	637.54	48.50	309.21	328.33	8	0.79	244.09
2017	38.25	660.00	48.50	320.10	339.90	9	0.77	245.33
2018	0.00	660.00	3.50	23.10	636.90	10	0.74	17.19
2019	0.00	660.00	3.50	23.10	636.90	11	0.72	16.69
2020	0.00	660.00	3.50	23.10	636.90	12	0.70	16.20
2021	0.00	660.00	11.30	74.58	585.42	13	0.68	50.79
⋮	⋮	⋮	⋮	⋮	⋮	⋮	⋮	⋮
2106	0.00	660.00	3.50	23.10	636.90	98	0.06	1.28
2107	0.00	660.00	11.30	74.58	585.42	99	0.05	4.00
Sum				**12,745**	**51,911**			**4,142**

Notes
Baseline is assumed to have been reached at 660 returning fish; 3.0%: Discount rate; 6%: Annual recovery of population to baseline; 2008: Base year (the year the analysis occurs in); DASSYs: Discounted Atlantic salmon service years; Sums rounded
To shorten the table, some of the results were omitted and substituted by the ellipsis

Table 12.11 Water Abstraction, River Itchen—debit calculation: 2.5% annual years of recovery to baseline, Atlantic salmon, years 2017–2107

	(A)	(B)	(C)	(D)	(E)	(F)	(G)	(H)
Year	Recovery to baseline (number of fish)	Baseline +	Fish lost due to water abstraction (%)	Number of fish lost	Number of fish returning to river	Year, base year	Discount factor	DASSYs
	(A)	(B) = 2.50%	(C)	(D) = (B) × (C)	(E) = (B) − (D)	(F)	(G) = 3.00%	(H) = (D × G)
2008		400	11.30	45.20	354.80	0	1.00	45.20
2009	10.00	410.00	3.50	14.35	395.65	1	0.97	13.93
2010	10.25	420.25	48.50	203.82	216.43	2	0.94	192.12
2011	10.51	430.76	11.30	48.68	382.08	3	0.92	44.54
2012	10.77	441.53	11.30	49.89	391.63	4	0.89	44.33
2013	11.04	452.56	11.30	51.14	401.42	5	0.86	44.11
2014	11.31	463.88	48.50	224.98	238.90	6	0.84	188.42
2015	11.60	475.47	48.50	230.61	244.87	7	0.81	187.50
2016	11.89	487.36	48.50	236.37	250.99	8	0.79	186.59
2017	12.18	499.55	48.50	242.28	257.27	9	0.77	185.69
⋮	⋮	⋮	⋮	⋮	⋮	⋮	⋮	⋮
2106	0.00	660.00	3.50	23.10	636.90	98	0.06	1.28
2107	0.00	660.00	11.30	74.58	585.42	99	0.05	4.00
Sum				**12,358**	**50,654**			**3,841**

Notes

Baseline is assumed to have been reached at 660 returning fish; 3.0%: Discount rate; 2.5%: Annual recovery of population to baseline; 2008: Base year (the year the analysis occurs in); DASSYs: Discounted Atlantic salmon service years; Sums rounded

To shorten the table, some of the results were omitted and substituted by the ellipsis

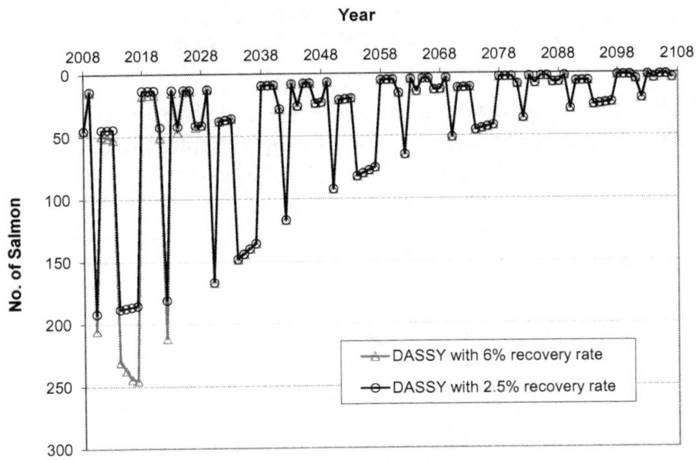

Fig. 12.9 Interim loss calculations (discounted Atlantic salmon service years) using 2.5 and 6% recovery rates to baseline

(benefits) transfer because a value estimate in the literature is transferred to the case study site and time period of the current analysis. The transfer could be unadjusted (using the same estimate found in the literature) or adjusted (adapting the estimate found in the literature to the factors at the case study site – as much as the data allow).

The most extensive database for economic value estimates that is publicly available online and that can potentially be used in this context is the Environmental Valuation Reference Inventory (EVRI).[3] As a searchable database of empirical studies, EVRI is a very useful source for a value transfer exercise.

We searched EVRI for this case study and found one study to be particularly relevant, namely, the report by Radford et al. (2001). The overall objective of that report was to estimate the total market value for inland fisheries in England. The estimate was part of a project aimed at determining the benefits or value provided by inland fisheries in order to inform policies in this area, notably regarding fishing rights.

One component of that study was an estimate of monetary value for each salmon caught in privately owned recreational inland fisheries in England.[4] The value was estimated through a hedonic pricing model, which seeks to establish how the market value of a private fishery varies with its characteristics, for example, its facilities (e.g., nearby parking), the population living in the surrounding area, and the number of salmon caught. The relationship between the number of salmon caught at a particular fishery and the value of a fishery is an indication of the value of the salmon population—the implicit price per salmon caught.

---

[3]www.evri.ca.

[4]Radford et al. (1991) note that almost all the inland fisheries in England are private properties and hence can be bought and sold on the market.

The market value of the fisheries was obtained through a survey of fishery owners, along with other information such as the number of salmon caught in the five preceding years and the other facilities available on the site. With that data, a statistical relationship between the market value of the property and its attributes was estimated to establish the effect of the number of salmon caught on the value of the property. This provided market price information on the value of the salmon population.

Thus, Radford et al. (2001) find the average value per salmon in fisheries in England to be £7,791 in 2001 prices, or £8,790 in 2007 prices. The authors note that this value is in line with earlier estimates by Radford et al. (1991). Note that the hedonic pricing methodology does not account for the non-use value of salmon and is therefore a lower bound of the total economic value.

In order to relate the number of fish lost each year due to the water abstraction scheme to an estimation of the implicit price per salmon caught, an estimate of the catch rate is needed, that is, the percentage of salmon population that is caught. This was done by compiling figures on the yearly salmon population and number of

Table 12.12 Water Abstraction, River Itchen—debit calculation: 6% annual recovery rate, Atlantic salmon, monetary value, years 2021–2106

Year	Number of fish lost	Catch rate (%)	Loss of salmon caught	Value per fish caught (£)	Discount factor	Discounted loss (£)
	(A)	(B)	(C) = (A) × (B)	(D)	(E)	(F) = (C) × (D) × (E)
2008	45.2	51	23.1	8,790	1.00	202,627
2009	14.8	51	7.6	8,790	0.97	64,589
2010	218.0	51	111.2	8,790	0.94	921,082
2011	53.8	51	27.5	8,790	0.92	220,853
2012	57.1	51	29.1	8,790	0.89	227,286
2013	60.5	51	30.8	8,790	0.86	233,906
2014	275.2	51	140.3	8,790	0.84	1,033,172
2015	291.7	51	148.8	8,790	0.81	1,063,264
2016	309.2	51	157.7	8,790	0.79	1,094,233
2017	320.1	51	163.3	8,790	0.77	1,099,790
2018	23.1	51	11.8	8,790	0.74	77,055
2019	23.1	51	11.8	8,790	0.72	74,810
2020	23.1	51	11.8	8,790	0.70	72,631
2021	74.6	51	38.0	8,790	0.68	227,666
⋮	⋮	⋮	⋮	⋮	⋮	⋮
2106	23.1	51	11.8	8,790	0.06	5,716
2107	74.6	51	38.0	8,790	0.05	17,918
Sum			12,745.5			**£18,569,352 (~ €25 million)**

Notes Column A is the same as Column D of Table 12.10
To shorten the table, some of the results were omitted and substituted by the ellipsis

Table 12.13 Water Abstraction, River Itchen—debit calculations, 2.5% annual rate of recovery, Atlantic Salmon monetary value, years 2017–2106

Year	Number of fish lost	Catch rate (%)	Loss of salmon caught	Value per fish caught (£)	Discount factor	Discounted loss (£)
	(A)	(B)	(C) = (A) × (B)	(D)	(E)	(F) = (C) × (D) × (E)
2008	45.2	51	23.1	8,790	1.00	202,627
2009	14.4	51	7.3	8,790	0.97	62,456
2010	203.8	51	103.9	8,790	0.94	861,260
2011	48.7	51	24.8	8,790	0.92	199,691
2012	49.9	51	25.4	8,790	0.89	198,721
2013	51.1	51	26.1	8,790	0.86	197,756
2014	225.0	51	114.7	8,790	0.84	844,657
2015	230.6	51	117.6	8,790	0.81	840,557
2016	236.4	51	120.5	8,790	0.79	836,477
2017	242.3	51	123.6	8,790	0.77	832,416
⋮	⋮	⋮	⋮	⋮	⋮	⋮
2106	23.1	51	11.8	8,790	0.06	5,716
2107	74.6	51	38.0	8,790	0.05	17,918
Sum			12,358.6			**£17,220,964 (~ €23million)**

Notes Column A is the same as Column D in Table 12.11
To shorten the table, some of the results were omitted and substituted by the ellipsis

salmon caught published by the EA (2006). The average catch rate over the period 1996–2006 was projected over the next 100 years. Of course, this is a simplification, and a more sophisticated approach could be used to reflect, for example, the impact of the size of the stock on the catch rate. Using the lost salmon estimates from Tables 12.10 and 12.11, Tables 12.12 and 12.13 provide annual breakdowns of the calculations that were used to obtain the value lost over 100 years with 6 and 2.5% recovery rates.

The first step was to estimate the number of salmon that cannot be caught at fisheries along the River Itchen as a result of the reduction in the salmon population (column C in Tables 12.12 and 12.13). Thereafter, the loss of salmon caught was multiplied by the value of the salmon and then discounted back to the base year. Finally, the annual losses were summed over 100 years to obtain the total monetary loss due to the water abstraction scheme.

Following this approach, the estimated monetary loss over 100 years, which is implied by the water abstraction scheme, is between £17 and £18.5 million for an assumed recovery of 6 and 2.5%, respectively.

12.4 Determining the Credits

12.4.1 Remediation Alternatives

The objective for the complementary remediation needed to balance the damage calculated in the previous section is determined by the HD. This must ensure that the overall Natura 2000 network is protected, as defined by Article 6(4) of the HD.

Damage to the *Ranunculus* habitat should be addressed by remediating the same habitat type—'like-for-like remediation.' Guidance from the European Commission (2007) states that compensatory measures can consist of:

- Recreating a habitat on a new or enlarged site, to be incorporated into Natura 2000;
- Improving habitat on part of the site or on another site, proportional to the loss due to the project; and
- In exceptional cases, proposing a new site under the HD.

Within the chalk river system of southern England, there are numerous rivers that have significant reaches that are damaged or degraded and are not part of Natura 2000. Indeed, The State of England's Chalk Rivers (Environment Agency 2004) states that 31% of chalk river sites monitored were in poor or very poor condition and 57% had been 'significantly modified or worse.' One option for remediation would be to restore these rivers such that they could be incorporated into Natura 2000. An alternative would be to undertake restoration on the River Itchen or another chalk river SAC in England that is 'equivalent to the loss' calculated in Sect. 12.3.

Simply designating a new chalk river SAC in its current state would not seem to represent any gain in biodiversity and would not be an addition to what the United Kingdom should be contributing to the Natura 2000 network as part of its responsibilities under the HD. In other words, designation of SAC would not generate additional credits.

A similar approach could be taken to remediate the Atlantic salmon population. It would be possible to improve salmon habitat on another chalk river in England or to enhance habitat on the River Itchen provided this work can be shown to be additional to what the United Kingdom would have had to contribute to be in compliance with the HD as for instance proposed by Holmes (2003).

12.4.2 Selecting Remediation Projects

Techniques for river restoration have been developed over recent decades throughout Europe and much of the world. These restoration techniques involve renaturalising rivers by removing impediments to natural processes of erosion and deposition, reconnecting lost meanders and braided channels, replacing natural riverine features such as woody debris, and recreating channel features such as pool-riffle sequences, gravel bars, and islands. These types of restoration schemes have been shown to have dramatic effects on the macroinvertebrate community, macrophyte growth, and fish populations, particularly on the spawning success of salmonid species.

The effects of such river restoration projects on chalk rivers and on the metrics used in this case study (macroinvertebrate and Atlantic salmon populations) has not been quantitatively monitored. However, there are some examples that can be used to scale the benefits of river restoration projects using these metrics.

To gain an understanding of the relative improvement (service gain) from river restoration on the invertebrate community, a reference site approach was taken. Data regarding a silted section of the River Itchen upstream of an impoundment were used to compare this section with similar reaches of the river where flow was good and the floating *Ranunculus* habitat was typical.

To evaluate improvements to salmon spawning habitat, data relating to restoration of two reaches of a headwater stream in the River Avon were obtained from the Environment Agency (2007).

12.4.3 Calculating the Credit

Macroinvertebrate community metric

The first step in remediation is to calculate the area of the length of river that would need to be restored in order to remediate the effects of water abstraction calculated in Sect. 12.3. It was first necessary to obtain information on the percentage service gain, measured in terms of both the abundance of key flow-dependent invertebrate species and the overall diversity of species within both a silty, degraded reach of the river and a healthy reach of the river. These were the two invertebrate metrics considered in Sect. 12.3. Data showing improvement differences between the invertebrate community in good-quality chalk river habitat and degraded chalk river habitat were obtained from the EA. Data showed a 90% difference in species abundance between the degraded and the healthy river sections. Species diversity indices showed a less dramatic change, with only a 16% increase in species diversity between degraded and good-quality habitats.

Although the percentage habitat damage (debit) using invertebrate abundance was large (71%) compared to the change in species diversity (12.5%), the amount of potential service gain from river restoration was roughly comparable (90 and 16%, respectively). Consequently, the area of habitat needed to be created in order to provide the necessary remediation (credit) over a 100-year period was not significantly different. Assuming a 90% service gain accumulates over a 5-year recovery period and 100 years of benefits, 1 ha of habitat restoration will provide just over 27 DSHaY. Assuming a 16% service gain accumulates over a 5-year recovery period and 100 years of benefits, 1 ha of habitat restoration will provide 4.88 DSHaY. Provision of 165 ha of habitat service years would require restoration of 165/4.88 = 33.8 ha of river in present value terms.

Table 12.14 Water Abstraction, River Itchen—Calculation of increase in salmon fry density following river restoration

Sites	Density, salmon fry/100 m^2
Site 1	
2003	15.0
2006	10.7
2007	45.3
Site 2	
2003	10.8
2006	11.3
2007	128.4
Mean 2003	**12.9**
Mean 2006–2007	**48.9**
Increase in number of salmon fry	**36.0**
% increase	**279**

Atlantic salmon metric

Information on Atlantic salmon spawning success from the River Wyle, a tributary of the River Avon, was obtained before and after a 2003 river restoration project. The project consisted of two restored reaches of the river. Numbers of juvenile salmon (fry) were recorded in 2003, prior to restoration, and again in 2006 and 2007 (Table 12.14). Calculations of the increase in numbers of salmon fry from improved spawning success need to be translated into predicted numbers of returning adult fish because this was the metric used to calculate damages or debits.

Salmon are subject to significant levels of mortality at each stage of their life cycle. Some simple relationships between numbers of fry, par, smolts, and returning adults were calculated from the literature Baglinièrea et al. (2005), as follows:

- Fry—par: 50% survival;
- Par—smolt: 10% survival; and
- Smolt—returning adult: 5% survival.

From the increase of 36 fry/100 m^2 of river restoration, one can expect: $36 \times 0.5 \times 0.1 \times 0.05$ returning adults = 0.09 returning adults.

12.5 Scaling Remediation

12.5.1 Macroinvertebrate Community Metric

Given 27 DSHaYs per hectare of restoration (see Sect. 12.4.3), provision of 623 DSHaYs (see Table 12.6) would require the restoration of 623/27 = 23 ha of river restoration in present value terms. Assuming a river width of 10 m, this is equivalent to 23 km of river restoration (1 km of restoration = 1,000 m × 10 m = 10,000 m^2 = 1 ha).

Using the same assumptions about river width, the 16% service gain assumption (see Sect. 12.4.3) equates to 33.8 km of river restoration. Thus, despite the significant differences in the percentage losses and gains using the different invertebrate metrics, these balance each other out so that the area of habitat restoration needed is similar.

12.5.2 Atlantic Salmon Metric

The number of returning salmon needed to compensate for the damage caused by water abstraction was calculated in Sect. 12.3 using two rates of baseline recovery. The larger one (4,142 discounted Atlantic salmon service years from Table 12.10) was used for the purposes of this case.

Assuming a 6-year recovery period, 100 years of benefits, and a 3% discount rate, the 100 m^2 of river restoration will provide 2.63 DASSYs, as illustrated in Table 12.15. The rate of service gain from the river restoration project was assumed to provide increasing amounts of service (in terms of numbers of returning salmon) over the first six years following the restoration, with a 10% service gain in year 1, 25% in year 2, 50% in year 3, 70% in year 4, 90% in year 5, and 100% in year 6. In other words, the 100-m^2 area reaches its full capacity of facilitating returning fish (0.09 fish/100 m^2) by year 6, at 100% of service provision. Assuming a river width

Table 12.15 Water Abstraction, River Itchen—Calculation of credit over 100 years assuming a 6-year recovery period and 3% discount rate

Year	Year, base year	Discount factor	Returning fish (number/100 m^2)	Discounted credit (DASSYs per 100 m^2)
(A)	(B)	(C) = 3%	(D)	(E) = (C) × (D)
2009	1	0.97	0.009	0.087
2010	2	0.94	0.022	0.084
2011	3	0.92	0.045	0.082
2012	4	0.89	0.063	0.079
2013	5	0.86	0.081	0.077
2014	6	0.84	0.09	0.075
⋮	⋮	⋮	⋮	⋮
2104	96	0.06	0.09	0.0052
2105	97	0.06	0.09	0.0051
2106	98	0.06	0.09	0.0049
2107	99	0.05	0.09	0.0048
2108	100	0.05	0.09	0.0046
Total				**2.63**

Notes Provision of 4,142 DASSYs would require 4,142/2.63 × 100 m^2 = 15.75 ha of restored river

To shorten the table, some of the results were omitted and substituted by the ellipsis

of 10 m, this is equivalent to 15.75 km of river restoration (1 km of restoration = 1000 m × 10 m = 10,000 m^2 = 1 ha).

12.5.3 Consideration of Potential Remediation Projects

Ideally, river restoration takes place on the river that has been damaged, in this case the River Itchen. However, due to the quantity of remediation necessary (between 33.8 and 14.58 km), it may not be possible to identify a sufficiently degraded length in the river to provide sufficient remediation. This is particularly true where the slower-flowing, silty reaches of the river can provide important habitat for a number of typical chalk river species, for instance, juvenile stages of lamprey. Restoration projects must be sensitive to the need for the sufficient conservation of these slow-flowing reaches. Identification of appropriate remediation projects is further complicated by the need to identify projects that provide a substantial increase in service gain, that is, from a highly degraded condition to one of high ecological function. Given these constraints, it seems unlikely that sufficient restoration projects in the River Itchen alone would be identified. If this is the case, it would be necessary to identify one or more additional rivers in England on which to undertake restoration work. The geographical distance between the river(s) and the River Itchen may require addition of a displacement factor to the amount of remediation provided.

12.5.4 Cost of Remediation

Costs of river restoration projects, which are taken from the River Restoration Centre,[5] vary substantially. However, they are limited to the cost of implementing the construction work and do not include associated ancillary costs. Table 12.16 considers potential costs associated with restoration of a 1-km stretch of river. Costs were valid at the time of this case study in 2007.

Using the above unit costs, it was possible to calculate total cost of remediation using the different metrics as shown in Table 12.17. Costs for remediation of floating *Ranunculus* habitat range between approximately €10 and €15 million. The cost for remediation using the Atlantic salmon population as the metric is just over €7 million. These differences in remediation costs illustrate the importance of finding a metric that best reflects a true level of damage to the environment.

[5] www.therrc.co.uk.

Table 12.16 Water Abstraction, River Itchen—Costs of river restoration per kilometre

	£	€
Planning	15,000	21,429
Mobilization	5,000	7,143
Preliminary sampling	5,000	7,143
Implementation	150,000	214,286
Operations and management	50,000	71,429
Oversight by competent authority	15,000	21,429
Monitoring and reporting	25,000	35,714
Overhead	25,000	35,714
Contingency	25,000	35,714
Total	**315,000**	**450,000**

Table 12.17 Water Abstraction, River Itchen—Remediation costs for different lengths of river restoration

	Length of river restoration required (assuming 10-meter-wide river) (km)	Cost (€)
Cost using change in invertebrate species diversity as the metric	23.02	10,359,000
Cost using change in invertebrate abundance as the metric	33.8	15,210,000
Cost using returning Atlantic salmon as the metric	15.75	7,087,500

12.5.5 Value-to-Cost Equivalency Approach

The economic value of damage using the Atlantic salmon metric, as estimated in Sect. 12.3.2, is between €23 and €25 million in present value terms over 100 years of lost services (2007 exchange rate). This is in fact a conservative estimate in that it (a) is only the salmon angling value and hence excludes non-use values and (b) assumes that the unit economic value of salmon caught remains the same over 100 years. Despite this, the damage (debit) is significantly greater than the credit—at least €23 million compared to the cost of just over €7 million. Thus, if remediation using Atlantic salmon was claimed to be disproportionately costly, this comparison could be shown to prove the opposite.

12.6 Monitoring and Reporting

It is important that future monitoring demonstrates the necessary improvement of habitat required to provide sufficient remediation during the first five to six years of restoration implementation. Monitoring should be performed in order to record habitat characteristics, including invertebrate community and Atlantic salmon spawning productivity, in terms of numbers of fry produced.

12.7 Conclusions

This case study illustrates how equivalency analysis can be used to calculate the magnitude of compensation required in an *ex ante* case where damage to a Natura 2000 site is predicted to occur as a consequence of abstraction for public water supply. It is based on a fictitious scenario in which damage to the Natura 2000 site is permitted in accordance with Article 6(4) of the HD. It could also be an example of 'imminent threat' in the context of the ELD, given that if abstraction continues the damage is inevitable.

The magnitudes of damage (debit) and remediation (credit) were calculated using two metrics: changes to the floating *Ranunculus* habitat and Atlantic salmon. Changes in the aquatic macroinvertebrate community were shown to provide a sensitive measure of changes in the conservation status of the European-protected floating *Ranunculus* habitat. Choosing an appropriate method for measuring these changes so as to reflect changes in the quality of the habitat proved to be problematic. Changes in species diversity and in the abundance of key invertebrate groups were investigated as part of this case study. Change in species abundance appeared to provide a better measure of change due to the inherent resilience of faunal diversity to low-flow events.

A second metric, use of numbers of returning Atlantic salmon, was based on a salmon migration model developed by the EA. The result was a calculation of roughly half the quantity of remediation required than when the invertebrate metric was used. However, the calculation of benefits likely to accrue from river restoration schemes for salmon was based on a limited sample from a reach of river where salmon have habitually spawned. It is probably unrealistic to expect similar levels of benefits to arise from restoration of the entire length of river, which would be needed to remediate the calculated damages. In this case, considerably more remediation would be needed.

Although variable, the quantity of remediation required to offset predicted damage is considerable. It is very unlikely that sufficient length within the River Itchen could be restored. Consequently, additional river restoration projects on other similar chalk rivers in southern England would need to be identified. In many instances, these restored rivers would in turn need to be added to the Natura 2000 network. The cost of implementing the necessary river restoration is considerable but low compared to lower-bound estimates of damage of €23 million.

Appendix: Flow Thresholds Set with Reference to Local Investigations on the River Itchen—Summary

A trend linking invertebrate community variation and antecedent summer Q95 flow was identified. Based on multivariate ordination techniques, a statistically and ecologically significant community change was shown to occur as flows fell below 0.861–0.844 standardised flow units. Samples collected when flows were greater than or equal to 0.861 units contained typical chalk stream invertebrate communities, whereas those collected when flows were less than or equal to 0.844 units were already impacted. No samples were available when flows were between 0.861 and 0.844 standardised units. It is therefore not possible to be specific about the impacts of flow within this narrow range.

The community shift that occurs between 0.861 and 0.844 standardised flow units was evident at sites throughout the River Itchen catchment. The shift was primarily caused by a reduction in the abundance of macroinvertebrates that prefer fast-flowing water and are highly characteristic of the typical chalk stream community.

Figure 12.10 summarises the community change that occurs between 0.861 and 0.844 standardised flow units.

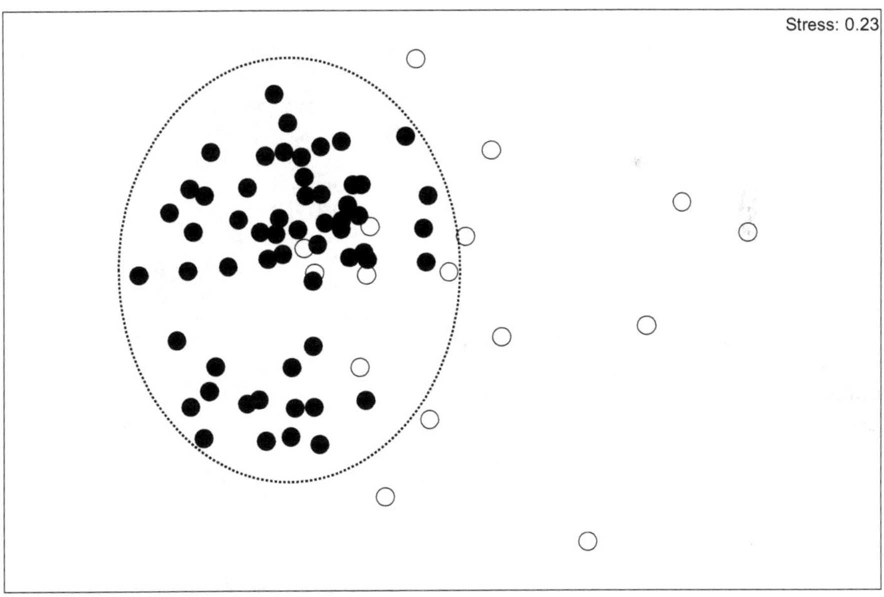

Fig. 12.10 Ordination of River Itchen samples highlighting samples (in black) collected when summer Q95 flow was greater than or equal to 0.861 standardised flow units, and samples (in white) collected when summer Q95 flows were less than or equal to 0.844 units. These sample groups were shown to be significantly different ($p = 0.001$) (*Source* Exley 2006)

References

Atkins. (2006). *River Itchen SAC review of consents, water resources stage 4 project, site action plan: Appendix B—Development of a stage 4 target flow regime*, Unpublished report to the Environment Agency.

Baglinièrea, J.-L., Marchandb, F., & Vauclin, (2005). Inter-annual changes in recruitment of the Atlantic salmon (Salmo salar) population in the River Oir (Lower Normandy, France): Relationships with spawners and in-stream habitat. *ICES Journal of Marine Science: Journal du Conseil, 62*(4), 695–707.

Environment Agency. (2004). *The State of England's Chalk Rivers. A report by the UK Biodiversity Action Plan Steering Group for Chalk Rivers*, Environment Agency and English Nature.

Environment Agency. (2006). *Salmon stock assessment report: Annual assessment of salmon stocks and fisheries in England and Wales*, Bristol.

Environment Agency. (2007). *Results of electro-fishing surveys, Langford Lakes, 2003–2007, River Wyle, Avon Catchment*, Unpublished data provided by the Environment Agency.

Environment Agency and CEFAS. (2006). *Annual assessment of salmon stocks and fisheries in England and Wales 2006*. Report by Environment Agency, Centre for Environment, Fisheries & Aquaculture Science.

European Commission. (2007). *Guidance document on Article 6(4) of the 'Habitats Directive' 92/43/EEC*, Brussels.

Exley, K. P. (2003). *The River Itchen macrophyte community*, Unpublished report to the River Itchen Sustainability Study, November.

Exley, K. P. (2004). *Macroinvertebrate habitat and community status of the River Itchen*, Unpublished report to the River Itchen Sustainability Study.

Exley, K. P. (2006). *River Itchen macroinvertebrate community relationship to river flow changes*, Unpublished Report, Environment Agency.

Extence, C. A. (1981). The effect of drought on benthic invertebrate communities in a lowland river. *Hydrobiol, 83,* 217–224.

Extence, C. A., Balbi, D. M., & Chadd, R. P. (1999). River flow indexing using British benthic macroinvertebrates: A framework for setting hydroecological objectives. *Regulated Rivers: Research & Management, 15,* 543–574.

Fewings, G. A. (2004). *Estimation of the effect of river discharge management on the passage of Atlantic salmon (Salmo salar L.) through the lower reaches of the River Itchen in Hampshire*, Unpublished report to the River Itchen Sustainability Study.

Holmes, N. (2003). *Concept designs for River Rehabilitation and enhanced Salmon Migration above Woodmill Pool*, Unpublished report to the River Itchen Sustainability Study.

Nijboer, R. C., Verdonschot, P. F. M., & Van Der Werf, D. C. (2005). The use of indicator taxa as representatives of communities in bioassessment. *Freshwater Biology, 50,* 1427–1440.

Radford, A. F., Hatcher, A., & Whitmarsh, D. (1991). *An economic evaluation of salmon fisheries in Great Britain*, a Research Report to the Ministry of Agriculture Fisheries and Food, Research Report 16, Centre for the Economics and Management of Aquatic Resources, University of Portsmouth.

Radford, A. F., Riddington, G., & Tingley, D. (2001). *Economic evaluation of inland fisheries: Economic evaluation of fishing rights*, R&D Technical Report/Project Record W2-039/TR/1 (Module A), Report to the Environmental Agency for England and Wales, Bristol.

Riley, W. D., Mason, C., Rowlatt, S. M., Maxwell, D., Campbell, S., & Hull, S., (1998). *The efficacy of River channel modification in maintaining improvements in salmonid spawning gravels following cleaning: Final report*. CEFAS—contract report CO224, 169.

River Itchen Study Steering Group. (2004). *River Itchen sustainability study*, final technical report, by Halcrow Group Ltd, Wiltshire, England.

Scott, A., & Beaumont, W. R. C. (1993). *Improving the survival rates of Atlantic Salmon (Salmo salar L.) embryos in a chalk stream*. Institute of Fisheries Management, Annual Study Course: Cardiff.

Solomon, D. J. (2002). *Dibden terminal inquiry—Proof of evidence of David Solomon Topic: Fisheries and Fish*, 55.

Solomon, D. J. (2004). *Development of action profiles for salmon, River Itchen sustainability study technical Ref: E(AP)02*, Report to the Environment Agency.

Chapter 13
Calculating Damage to Alpine Brown Trout Using Equivalency Analysis

Jamie V. Holmes and Joshua Lipton

Abstract This hypothetical case study presents a method of calculating environmental damage to brown trout caused by hard rock mine pollution in an alpine river. In this scenario, heavy metals are released into a river as a consequence of the failure of a tailings dam impoundment at a hard rock mine, with subsequent adverse impacts to a brown trout fishery through impairment of water and sediment quality. This hypothetical tailings dam failure is assumed to have occurred in May 2014, with subsequent fish population studies conducted in 2014, 2015, and 2016. Brown trout populations recover slowly after the incident and, through linear extrapolation, we assumed that the impacted river would return to pre-release baseline trout populations by 2030. Primary remediation actions were assumed to include repairing the tailings dam and removing tailings from proximal areas of stream channel and floodplain, but interim losses were assumed to occur prior to and following implementation of the primary remediation. To assess environmental damages for these interim losses, we used an equivalency analysis with trout density as the metric to quantify damages and remediation credits. In-stream habitat enhancement was selected as the preferred remediation approach to compensate for interim loss. Habitat enhancements that would increase trout density include emplacement of structures such as large woody debris or boulders, plunge pool construction, and riparian revegetation. We estimated post-remediation increases in trout densities and quantified the credit per hectare of stream remediation. Finally, we calculated the amount of remediation (number of hectares) and associated potential costs required to offset the damage.

Keywords Resource equivalency analysis · Habitat equivalency analysis Alpine habitat · Brown trout · Mine pollution

---

J. V. Holmes (✉) · J. Lipton
Abt Associates Inc, 1881 Ninth St., Suite 201, Boulder, CO 80302, USA
e-mail: Jamie_Holmes@abtassoc.com

J. Lipton
e-mail: Josh_Lipton@abtassoc.com

J. Lipton et al. (eds.), *Equivalency Methods for Environmental Liability*,
https://doi.org/10.1007/978-90-481-9812-2_13

13.1 Introduction

This hypothetical case study addresses a scenario in which heavy metals released from a mine contaminated an alpine river, resulting in damages to aquatic habitat and resident populations of brown trout (*Salmo trutta*). We used an equivalency analysis approach to calculate liability for environmental damage based on changes to brown trout densities.

This hypothetical case study is organised as follows. First, we describe the scenario, including the timing of the spill event, the response actions, and the adverse impacts to brown trout populations. We then discuss an equivalency analysis framework in which the metric for loss is based on brown trout density. We determine and quantify debits and credits using this metric to scale compensatory remediation actions focused on aquatic habitat enhancement projects.

Because this hypothetical scenario deals with environmental damage that has already happened ('*ex post*' scenario), equivalency methods are applied under Annex II of the Environmental Liability Directive (ELD). The ELD does not apply retroactively to any incident or damage that occurred before 30 April 2007 and hence we assume that the mining incident occurred after that date.

13.2 Description of the Incident

The scenario used for this hypothetical case study focuses on a hard-rock mine in an alpine setting. The mine extracts and crushes mined ore, concentrates the metals on-site, and places the waste material generated after processing, known as tailings, into a dammed impoundment near the headwaters of a river. This is a common practice worldwide in the hard-rock mining industry.

Prior to the tailings release, the river downstream of the mine supported a healthy, self-sustaining brown trout fishery. Adverse impacts from other land uses (e.g., transportation corridors, logging) were minimal, providing the trout with quality habitat features such as overhanging riparian vegetation and large structures such as boulders and woody debris that create habitat diversity such as deep plunge pools. Thus, we assumed that the river contained natural habitats listed in Annex I of the Habitats Directive (HD) and that any adverse impacts to the river would affect protected fauna species listed in Annex II of the HD.

In our scenario, the tailings dam was breached in May 2014, releasing tailings into the river. Specifically:

- Several tonnes of tailings were released from the tailings impoundment into the river;
- The cognizant mining company quickly repaired the breach to prevent additional tailings releases; and
- Over the course of the next several months, the mining company removed much of the tailings from the stream bed and adjacent floodplain close to the impoundment.

We assumed that such an uncontrolled release of mine tailings into a mountain catchment resulted in a large pulse of heavy metals into the water. At elevated concentrations, metals are toxic to brown trout (e.g., Farag et al. 1995; Woodward et al. 1995a, b; Clements and Rees 1997; Brinkman and Hansen 2007). In addition, fine-grained tailings can smother trout spawning gravels, stress streamside vegetation, and harm invertebrates that support the aquatic food chain.

Our hypothetical spill scenario assumes that many brown trout are killed in the vicinity of the mine. Authorities mobilise a response team to monitor brown trout populations for several kilometres downstream of the site. The response team also measures trout populations in an adjacent reference watershed with similar land use but no mine waste. This allows for trout density comparisons to the impacted river. Although comparison to trout populations upstream of the mine release are an alternative approach to evaluating baseline conditions, mining operations are frequently at or near the headwaters of a drainage, which limits comparisons to upstream fish populations. Therefore, our scenario adopts a reference site approach to baseline assessment. Each 2014 monitoring location is re-examined at the same time of year in 2015 and 2016, providing data to estimate the rate of brown trout recovery.

Although this scenario is hypothetical, we have conducted natural resource damage assessments at several mine sites in the United States where releases have had similar adverse effects on aquatic habitat. The techniques used to quantify the damages in this example thus are similar to techniques used in actual damage assessments in the United States.

13.2.1 Affected Environments

For simplicity, we assumed that the impacts were restricted to shallow alpine aquatic habitat below tree line. The stream channel is relatively narrow (10–15 m wide) and shallow (less than 1 m deep at most times of year), with diverse channel morphology. Mature riparian vegetation comprising mixed grasses, shrubs, and trees provide bank stabilisation and a closed overhead canopy. Tree trunks and root wads (i.e., large woody debris) and boulders provide important habitat structure. These alpine streams typically have high gradients, with primarily gravel and cobble substrate and sequences of riffles, runs, and pools. Within the aquatic habitat, we would expect a diverse assemblage of macroinvertebrates, including ephemeropterans, plecopterans, and trichopterans, which provide a food base for the brown trout and other fish species, which in turn may serve as prey for birds and mammals.

In this hypothetical case study, we assumed that the condition of the aquatic habitat prior to the tailings dam breach was good but not pristine. Alpine mine sites typically host transportation infrastructure such as roads and railroads that may degrade trout habitat. To account for baseline degradation of the trout habitat prior to the release, we assumed that an adjacent watershed contains similar partially disturbed alpine habitat, but no tailings spill.

13.2.2 Primary Remediation Undertaken

In our scenario, the cognizant mining company responded quickly to the breach. Typical response actions would be to first repair the breach in the tailings dam, usually by pushing clean fill into the breach area with a bulldozer. Large deposits of tailings in the river and adjacent floodplain might be removed with heavy equipment, depending on access, risks, and the potential for collateral damage to human health and the environment. Even assuming removal of bulk tailings materials, when tonnes of fine-grained tailings are released into a river, those tailings likely would be deposited over many kilometres downstream and it would not be feasible to remove all metal sources after such a release. Thus, we assumed that heavy metal concentrations will remain elevated for years after the release, slowing the natural recovery of trout populations.

13.2.3 Potentially Affected Services

Although such a tailings spill would likely affect a variety of ecological and economic services, including habitat for invertebrates, fish, riparian vegetation, and potentially birds and mammals, our case study focuses on brown trout as representative indicators of aquatic damages. In an actual case, Competent Authorities might choose to look at multiple indicators of environmental harm and quantify the amount of remediation required to compensate for different indicators and metrics.

For our brown trout illustration, we assume that primary affected services would include:

- Fish habitat, including nutrient cycling, food resources, and spawning gravels;
- Food resources for piscivorous birds and mammals; and
- Recreational fishing.

13.3 Quantifying Debits

We used an equivalency analysis to quantify damages to aquatic resources, with brown trout density used as the metric used to assess service losses and gains. We could refer to this as a resource equivalency analysis (REA) approach because we are scaling remediation based solely on damages to brown trout resources. We could also call this a habitat equivalency analysis (HEA) approach, where habitat services are based solely on brown trout resources, which are an indicator of aquatic habitat health. In both cases, we are scaling remediation based on a single resource. In a REA framework, the calculation would typically include the number of trout lost (debit) and the number gained through remediation (credit). In the HEA framework, we would use the same data but calculate trout lost and gained per unit area of habitat. Our calculations in this hypothetical example are in units of habitat. However, we note that this approach does not attempt to quantify the effects of the spill on all potentially affected habitat services; rather, it is resource-to-resource scaling based solely on brown trout, with calculations in units of habitat rather than units of fish.

13.3.1 Baseline Conditions

We assumed that the impacted river supported a healthy brown trout fishery prior to the tailings spill. Because our chosen metric for evaluating damages was trout population density, brown trout populations prior to the spill would represent pre-incident baseline conditions. Although it would have been ideal to have measured trout population density immediately prior to the release, this situation rarely occurs. More commonly, pre-incident baseline trout density is established by measuring trout populations at a 'control' or 'reference' site, such as upstream of the release (if the mine is not at the headwaters), or in a nearby watershed with similar characteristics but without mine waste. Data would be collected at the reference and impacted locations using similar sampling methods and at similar times of year to minimize seasonal or weather-related variables.

13.3.2 Quantification of Losses

Trout populations are often quantified in terms of the number of fish per unit area, such as the number of trout per square meter of river. In our scenario, fisheries biologists conducted multiple pass depletion electrofishing to determine trout densities at sample sites downstream of the mine, as well as in a nearby reference river. Electrofishing was conducted in 2014 after the spill, then repeated in 2015 and 2016 at the same time of year to track trout recovery.

In this scenario, there were no brown trout in 2014 in the river reach closest to the tailings dam breach. In a reach further downstream, brown trout were present but at densities well below the baseline density found at the control site. We designated sampling sites with no trout to occur within river Reach 1, the most upstream reach. Sample sites with reduced trout densities were designated as river Reach 2, the more downstream reach. We assumed that Reach 1 is 3 km in length, with an average width of 10 m. Reach 2 is 5 km in length, with an average width of 15 m. Thus, the area of aquatic habitat in Reach 1 is 3.0 ha, and the area of aquatic habitat in Reach 2 is 7.5 ha.

The trout density measured at the reference site was 0.2 trout per m^2, which for ease of presentation we restated as 20 trout per 100 m^2. This baseline density did not change when the sites were resampled in 2015 and 2016. After the tailings spill

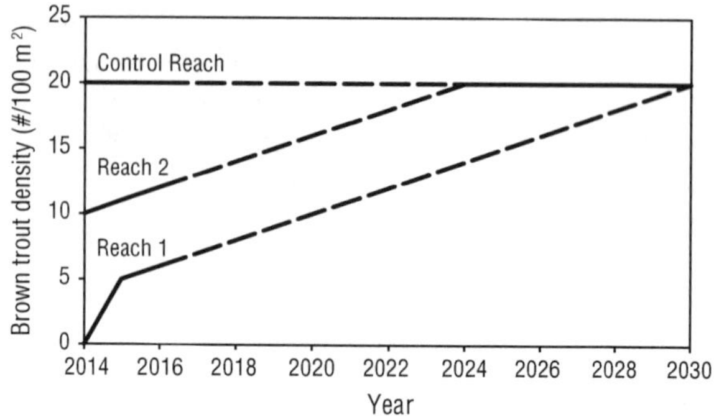

Fig. 13.1 Trout densities in the control reach and two impacted reaches over time (The dashed lines are estimates of future densities, based on linear interpolation of 2015 and 2016 data)

in 2014, the sample sites in Reach 1 had 0 trout per 100 m^2, and the average trout density at the sites in Reach 2 was 10 trout per 100 m^2, or half of the baseline density. In 2015, trout density in Reach 1 rebounded to 5 trout per 100 m^2, and the trout density in Reach 2 increased slightly to 11 trout per 100 m^2. Both reaches increased by an additional 5% in 2016, with 6 trout per 100 m^2 in Reach 1, and 12 trout per 100 m^2 in Reach 2 (Fig. 13.1). We assumed that trout densities in the impacted reaches would continue to increase at a rate of 5% per year, with Reach 2 achieving the baseline density of 20 trout per 100 m^2 in 2024 and Reach 1 in 2030 (Fig. 13.1).[1]

For this simplified case study, because habitat improvements were selected as the preferred remediation alternative (see below), we used relative brown trout density as the metric for proportional loss of aquatic habitat services such that the damage to aquatic habitat quality was assumed to be equal to the proportional decrease in trout density relative to baseline conditions. Thus, in 2014, there was a 100% loss of trout density services in Reach 1 and a 50% loss of services in Reach 2. In 2015, Reach 1 rebounded to 5 trout per 100 m^2 compared to a baseline density of 20, and thus service loss decreased from 100 to 75%. Trout populations in Reach 2 increased slightly from 10 to 11 trout per 100 m^2, and trout density service loss decreased from 50 to 45%. Because the hypothetical trout densities increased by 5% each subsequent year, trout density service loss decreased by 5% each subsequent year (Fig. 13.2).

We quantified debits using this approach using a base year of 2017 and a 3% discount rate. For each year that trout density was less than baseline, we multiplied the present value or discount factor by the percent loss of trout population and the area of the impacted habitat. This provided us with a debit for each year, expressed

[1]For simplicity of presentation, our illustration focuses on the total density of brown trout. In an actual case, it might be more appropriate to differentiate baseline and damaged trout populations by age/size structure, rather than a pooled density number. When performing equivalency analyses across age-structured populations, debits and credits could be tracked for individual ages/lifestages (e.g., trout fry, juveniles, adults), or could be converted to a single, normalized lifestage-equivalents (e.g., age-1 equivalents) using population life-tables.

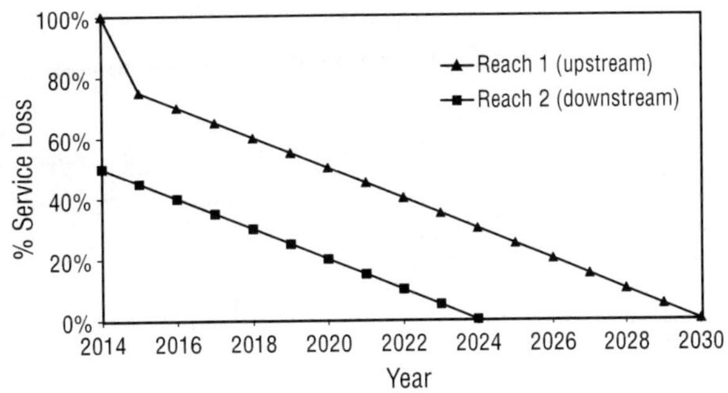

Fig. 13.2 Trout density service loss over time for Reach 1 and Reach 2 [Because habitat services are based on trout populations, service loss over time is the inverse of trout density over time (Fig. 13.1)]

in units of discounted service hectares, where the 'service' metric is based on a single resource, brown trout. We summed the discounted service hectares for each year of impact, providing a total debit in Discounted Service Hectare Years (DSHaYs). The debit for Reach 1 is 20.0 DSHAYs (Table 13.1) and the debit for Reach 2 is 20.7 DSHaYs (Table 13.2). Because we assumed that Reach 1 and Reach 2 were in the same type of aquatic habitat, we added the debit for each reach, for a total loss of 40.7 DSHaYs.

Table 13.1 Calculated debit for Reach 1 (upstream)

Year	Discount factor	Service loss (% reduction in brown trout density) (%)	Area (ha)	Present value debit (DSHaYs)
2014	1.09	100	3.0	3.28
2015	1.06	75	3.0	2.39
2016	1.03	70	3.0	2.16
2017	1.00	65	3.0	1.95
2018	0.97	60	3.0	1.75
2019	0.94	55	3.0	1.56
2020	0.92	50	3.0	1.37
2021	0.89	45	3.0	1.20
2022	0.86	40	3.0	1.04
2023	0.84	35	3.0	0.88
2024	0.81	30	3.0	0.73
2025	0.79	25	3.0	0.59
2026	0.77	20	3.0	0.46
2027	0.74	15	3.0	0.33
2028	0.72	10	3.0	0.22
2029	0.70	5	3.0	0.11
2030	0.68	0	3.0	0.00
Total				**20.0**

Table 13.2 Calculated debit for Reach 2 (downstream)

Year	Discount factor	Service loss (% reduction in brown trout density) (%)	Area (ha)	Present value debit (DSHaYs)
2014	1.09	50	7.5	4.10
2015	1.06	45	7.5	3.58
2016	1.03	40	7.5	3.09
2017	1.00	35	7.5	2.63
2018	0.97	30	7.5	2.18
2019	0.94	25	7.5	1.77
2020	0.92	20	7.5	1.37
2021	0.89	15	7.5	1.00
2022	0.86	10	7.5	0.65
2023	0.84	5	7.5	0.31
2024	0.81	0	7.5	0.00
Total				**20.7**

13.4 Evaluating Potential Remediation Projects

We evaluated several types of potential remediation projects focused on enhancing trout populations, including:

- Reintroduction of fish through stocking;
- Angling management;
- Remediation of fish spawning habitats;
- Removal of abandoned transportation infrastructure (road and railroad beds) that impair existing trout habitat;
- Other improvements to habitat in the stream channel, such creation of side channels, islands, or in-stream pools, and addition of large woody debris or rocks to enhance habitat diversity; and
- Construction of vegetation buffer zones to reduce sediment inputs from transportation corridors and logging areas.

We evaluated these project types against the criteria presented in Chap. 5 that are relevant to this hypothetical case study (Table 13.3). Based on the project evaluation process, we determined that channel improvements represented the preferred remediation alternative.

13.5 Calculating Project Benefits and Scaling Compensatory Remediation

Because our scaling metric is brown trout density, we estimated the anticipated increase in brown trout density after completion of remediation, as well as the amount of time required for the benefits to brown trout to be realised.

Table 13.3 Evaluation of potential aquatic remediation projects

Evaluation criterion	Remediation project type					
	Stocking	Angling management	Spawning habitat	Road bed removal	Channel improvement	Vegetation buffers
Initial screening						
Nexus to damaged resources	+	+	+	+	+	+
Legal	+	+	+	+	+	+
No risks to public	+	+	+	+	+	+
Feasible	+	+	+	+	+	+
Don't cause harm	+	+	+	+	+	+
Public acceptance	+	?	+	+	+	+
Detailed screening						
Address priority resources	+	+	+	+	+	+
Use reliable methods	+	+	+	+	+	+
Costs < benefits	+	+	+	+	+	+
Readily scaled to debits	Low	Low	Low	Low	High	Low
Benefits can be quantified	Low	Low	High	High	High	High
Consistent with regional planning	High	Low	High	Low	High	High
Provide benefits to multiple resources	Low	High	High	High	High	High
Enhance public enjoyment	High	Low	High	High	High	High
Provide long-term benefits	High	?	High	High	High	High
Provide benefits to comprehensive segment of population	High	Low	High	High	High	High

We applied either +/− or high/low rankings to habitat-relevant criteria presented in Chap. 5
?: stands for unknown

If a specific location for habitat remediation is targeted, one could collect trout population data at the site and compare it to reference data to estimate the anticipated improvement in trout density. However, it is common for equivalency analysis calculations to be performed before specific projects are identified, requiring an alternative approach to estimate future service gains from remediation.

Often, one can rely on regional fish population data, where the data were collected using standardised multiple pass depletion electrofishing techniques. These data provide a range of trout densities across many rivers with different land uses and different habitat qualities. Using these data, we can estimate a typical increase in trout density after remediation in a yet-to-be-determined location.

If habitat quality data are available for each trout sampling site in the regional fish population database, the data might be sorted by habitat, which would allow for calculation of average trout density in habitats of differential qualities. Lacking habitat data, one might evaluate the dataset as a whole and decide, for example, that the 80th percentile trout density across all locations is representative of a typical density in good habitat and that the 20th percentile trout density is representative of a typical density in poor habitat. One might then assume that a typical trout habitat remediation project could, for example, increase trout density from the 20th to the 80th percentile.

For this hypothetical scenario, we assumed that regional brown trout population data were available, that the data were collected using comparable methods (and thus comparisons of trout densities between locations were valid), and that habitat quality data were also available. The average brown trout density in high quality trout habitat across the entire region was 25 trout per 100 m^2. At sites where channel straightening, erosion, sedimentation, and poor riparian vegetation degraded trout habitat, the average trout density was 5 trout per 100 m^2.

Thus, we set 25 trout per 100 m^2 as the typical density for brown trout in high quality aquatic habitat, and we assumed that in-stream habitat remediation implemented at locations with poor habitat could increase brown trout density from 5 to 25 trout per 100 m^2. On a percentage basis for purposes of scaling remediation, this represents an increase in trout density from 20 to 100% of baseline trout density for high quality habitat (Fig. 13.3). For purposes of expressing credits from habitat

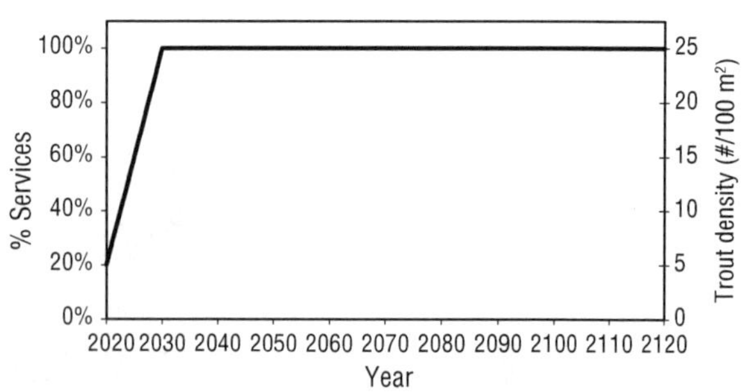

Fig. 13.3 Assumed increase in trout density and corresponding increase in services from in-stream habitat remediation projects (Because trout density is the metric for estimating services, the lines are identical)

enhancements on an area basis, the remediation projects thus would result in an 80% net increase in trout density services for each hectare of remediation.

Stream channel remediation can be completed relatively quickly. However, it takes several years for a remediated reach to develop natural habitat complexity and for brown trout to subsequently repopulate the reach. For example, if the degraded habitat naturally has a tree canopy, providing shade and woody debris, and those trees have been removed, it may take decades to reestablish the canopy. For this example, we assumed that improvements in brown trout densities could be achieved based on in-stream habitat improvements without waiting for reestablishment of a mature tree canopy. Specifically, we estimated that after remediation is complete, it would take 10 years, encompassing several generations of spawning brown trout, before brown trout densities reach 25 per 100 m^2 and the full benefit of the remediation is realised (Fig. 13.3).

To calculate credit, we again used a base year of 2017 and a 3% discount rate. We assumed that the remediation projects would be implemented in 2020, benefits would begin to accrue in 2021, and those benefits would continue to accrue through 2120 (after which benefits become negligible because of discounting). Trout densities increase linearly from 20 to 100% of baseline over a 10-year period (Fig. 13.3). For each year where trout density is higher than it would be without remediation, we multiplied the discount factor with the net percent service gain, providing us the credit for each year. We summed the discounted service hectares for each year of habitat improvement through 2120, providing total credit per hectare of remediation in DSHaYs. Using this method, the total unit credit is 20.2 DSHaYs per hectare of channel remediation (Table 13.4).

The total debit for the two impacted reaches combined (Tables 13.1 and 13.2) is 40.7 DSHaYs. With stream channel remediation providing 20.2 DSHaYs of credit per hectare (Table 13.4), the total amount of remediation required, rounded to a whole number, is 40.7/20.2 = 2 ha.

13.6 Calculating Total Liabilities

The cost of in-stream channel improvements for fish habitat can vary widely, depending on the extent and location of the proposed remediation. In the Vistula River example from Poland (see Chap. 9), it was estimated that channel remediation could be accomplished for €31,200/ha. This likely underestimates the remediation cost in western European Union countries and in the alpine areas described in this scenario. Table 13.5 presents an alternative cost scenario, based on alpine trout habitat remediation costs from our projects in the United States.

The costs presented in Table 13.5 include river/stream diversion to allow construction vehicle access, regrading of the stream channel, emplacement of habitat structures such as boulders and large woody debris, and revegetation of the stream bank. Design and oversight of the project are 20% of the construction cost, and mobilisation, demobilisation, and permitting are 10% of the construction cost. We

Table 13.4 Calculated HEA credit per hectare of stream channel remediation

Year	Discount factor	Services (% increase in brown trout density)	Net Service Gain (%)	Present value debit (DSHaYs)
2020	0.92	20	0	0.00
2021	0.89	28	8	0.07
2022	0.86	36	16	0.14
2023	0.84	44	24	0.20
2024	0.81	52	32	0.26
2025	0.79	60	40	0.32
2026	0.77	68	48	0.37
2027	0.74	76	56	0.42
2028	0.72	84	64	0.46
2029	0.70	92	72	0.50
2030	0.68	100	80	0.54
2031	0.66	100	80	0.53
2032	0.64	100	80	0.51
⋮	⋮	⋮	⋮	⋮
2119	0.05	100	80	0.04
2120	0.05	100	80	0.04
Total				**20.2**

Note To shorten the table, some of the results were omitted and substituted by the ellipsis

did not include land acquisition costs. We included long-term monitoring costs, but we did not include long-term maintenance, as it is generally not required for these types of projects. Assuming a 20% contingency, the total estimated cost for channel remediation is €427,000/ha (Table 13.5).

With a total of 2 ha of remediation required, the total estimated liability for this tailings spill scenario is €427,000 × 2 = €854,000 (Table 13.6).

Table 13.5 Estimated costs of channel remediation to improve brown trout habitat in alpine areas

Cost element	Estimated cost (€ per ha)
Construction	270,000
Design and oversight	54,000
Permitting, mobilization, and demobilization	27,000
Monitoring	5,000
Subtotal	356,000
Contingency (20%)	71,000
Total	**427,000**

Table 13.6 Total estimated liability

Debit (DSHaYs)	Credit (DSHaYs/ha)	Remediation required (ha)	Unit cost (€/ha)	Total cost (€)
40.7	20.2	2.0	427,000	854,000

Numbers are rounded

13.7 Conclusions

We presented a hypothetical damage scenario in which tailings from hard-rock mining were released into an alpine river that supports a population of brown trout. Brown trout were entirely eliminated from a 3-km reach of river near the tailings spill, and were decreased over an additional 5-km reach downstream. The mining company repaired the breach in the tailings dam and removed tailings from the stream channel where practicable. However, several years were required for trout populations to return to their pre-impact baseline density.

We performed an equivalency analysis to illustrate how interim losses could be calculated using trout density as the metric for evaluating service losses and gains. We assumed that the annual changes in brown trout densities for the years immediately following the spill are indicative of the rate of recovery, and we extrapolated the impacts to trout populations in the future based on those recovery rates.

Compensatory remediation projects were selected to provide benefits to alpine aquatic habitat. After applying a series of project evaluation criteria, we selected in-stream channel improvements as the preferred remediation alternative. We estimated that 2 ha of aquatic habitat would need to be remediated at a total cost of €854,000.

This hypothetical case study demonstrates how environmental damages associated with impacts to brown trout habitat can be calculated. The method presented herein is based on methods that we have used at similar sites in the United States.

References

Brinkman, S. F., & Hansen, D. L. (2007). Toxicity of cadmium to early life stages of brown trout (*Salmo trutta*) at multiple water hardnesses. *Environmental Toxicology and Chemistry, 26*(8), 1666–1671.

Clements, W. H., & Rees, D. E. (1997). Effects of heavy metals on prey abundance, feeding habits, and metal uptake of brown trout in the Arkansas River, Colorado. *Transactions of the American Fisheries Society, 126,* 774–785.

Farag, A. M., Stansbury, M. A., Hogstrand, C., MacConnell, E., & Bergman, H. L. (1995). The physiological impairment of free-ranging brown trout exposed to metals in the Clark Fork River, Montana. *Canadian Journal of Fisheries and Aquatic Sciences, 52,* 2038–2050.

Woodward, D. F., Hansen, J. A., Bergman, H. L., Little, E. E., & DeLonay, A. J. (1995a). Brown trout avoidance of metals in water characteristic of the Clark Fork River, Montana. *Canadian Journal of Fisheries and Aquatic Sciences, 52,* 2031–2037.

Woodward, D. F., Farag, A. M., Bergman, H. L., DeLonay, A. J., Little, E. E., Smith, C. E., et al. (1995b). Metals-contaminated benthic invertebrates in the Clark Fork River, Montana: effects on age-0 brown trout and rainbow trout. *Canadian Journal of Fisheries and Aquatic Sciences, 52,* 1994–2004.